The Roswell Crash has had a massive hypnotic effect on the collective psyche of Mankind ever since the U.S. Army issued a press release that a flying disc crashed in the New Mexican desert in July of 1947. Retracting their statement a day later, they have since orchestrated a massive psychological operation against the public to hide a long history of war crimes involving some of the most iconic of American heroes. At the core of this operation is the Government's most guarded secret of World War II: the Japanese Emperor's direct role in prosecuting the war.

Horrified upon learning of a published agenda to destroy *the Yellow Race, a young Hirohito fervently studied to become a preeminent Doctor of Marine Biology in order to create biocidal weapons of mass destruction and so defend his Empire. After successfully testing his genocidal weapons in the Chinese war, they were delivered to the Americans three days after Hiroshima in huge super-dirigibles equipped with planes and tanks. Realizing the potency of these weapons and their own potential doom, the Americans began a series of back-channel negotiations with Hirohito during which they acquiesced to his demands that he remain as Emperor and that America open up their markets to his country.*

This hidden history includes the revelations that the *Emperor's super-dirigibles were the source of the Battle of Los Angeles in February of 1942 and the infamous Roswell Crash of 1947. In an involved and labyrinthine tale that involves hidden hordes of gold, cut-throat politics, hidden technology and massive psychological manipulation of the masses,* ***The Roswell Deception*** *delivers the long anticipated "Disclosure" that humanity has been waiting for.*

Donald Trump, Jr. conducted an interview with his father, President Donald Trump, and he asked him the following question in June of 2020:

"Before you leave office, will you let us know if there's aliens? 'Cause this is the only thing I really want to know. I want to know what's going on. Would you ever open up Roswell and let us know what's really going on there?"

"There are millions and millions of people that want to go there, that want to see it. I won't talk to you about what I know about it, but it's very interesting," said President Trump. "Roswell's a very interesting place with a lot of people that would like to know what's going on."

THE ROSWELL DECEPTION AND THE DEMYSTIFICATION OF WORLD WAR II

BY DOUGLAS DIETRICH WITH PETER MOON

SkyBooks

NEW YORK

The Roswell Deception and the Demystification of World War II
Copyright © 2021 by Douglas Dietrich and Peter Moon
First printing, April 28, 2021

Cover art and illustration by Creative Circle Inc.
Typography by Creative Circle Inc.
Published by: Sky Books
   Box 769
   Westbury, New York 11590
   email: skybooks@yahoo.com
   website: www.skybooksusa.com

Printed and bound in the United States of America. All rights reserved. No part of this book may be reproduced in any form or by any electronic or mechanical means including information storage and retrieval systems without permission in writing from the publisher. This book includes photographs and quotations that are used in conjunction with Fair Use Copyright Law.

Library of Congress Cataloging-in-Publication Data

Douglas Dietrich / Moon, Peter
   The Roswell Deception and the Demystification of World War II
by Douglas Dietrich with Peter Moon
   312 pages
   ISBN 978-1-937859-23-7
1. World History 2. United States History
Library of Congress Control Number 2021933179

This book is dedicated to my mother, Dianna Sūjĭn-Lín Dietrich, and my father, George Joseph Henry Dietrich.

Other titles from
Sky Books

by Preston Nichols and Peter Moon
The Montauk Project: Experiments in Time
The Montauk Project Silver Anniversary Edition
Montauk Revisited: Adventures in Synchronicity
Pyramids of Montauk: Explorations in Consciousness
Encounter in the Pleiades: An Inside Look at UFOs
The Music of Time

by Peter Moon
The Black Sun: Montauk's Nazi-Tibetan Connection
Synchronicity and the Seventh Seal
The Montauk Book of the Dead
The Montauk Book of the Living
Spandau Mystery
The White Bat
L. Ron Hubbard — The Tao of Insanity

by Joseph Matheny with Peter Moon
Ong's Hat: The Beginning

by Stewart Swerdlow
Montauk: The Alien Connection
The Healer's Handbook: A Journey Into Hyperspace

by Alexandra Bruce
The Philadelphia Experiment Murder:
Parallel Universes and the Physics of Insanity

by Wade Gordon
The Brookhaven Connection

by Radu Cinamar
Transylvanian Sunrise
Transylvanian Moonrise
Mystery of Egypt — The First Tunnel
The Secret Parchment — Five Tibetan Initiation Techniques
Inside the Earth — The Second Tunnel
Forgotten Genesis
The Etheric Crystal — The Third Tunnel

CONTENTS

FOREWORD — by Peter Moon.................................9
INTRODUCTION — by Douglas Dietrich.................15
OPENING STATEMENT — by Douglas Dietrich........19
CHAPTER 1 — Japan, an Isolationist Empire..............25
CHAPTER 2 — Emperor Hirohito................................31
CHAPTER 3 — The American Army and Navy Flu....37
CHAPTER 4 — FDR..49
CHAPTER 5 — Communist Pedigree............................57
CHAPTER 6 — The Marine Corps...............................65
CHAPTER 7 — Drugs and War Mongering.................71
CHAPTER 8 — Pearl Harbor..83
CHAPTER 9 — The Chrysanthemum Throne.............89
CHAPTER 10 — The Chinese War...............................97
CHAPTER 11 — The United Nations........................113
CHAPTER 12 — The Battle of Los Angeles..............115
CHAPTER 13 — The Office of War Information.....127
CHAPTER 14 — Midway...137
CHAPTER 15 — Fu-Go Balloon Bombs....................141
CHAPTER 16 — The Triple Nickles Battalion..........149
CHAPTER 17 — Black Lives Didn't Matter.............153
CHAPTER 18 — Okinawa —Operations Detachment
                              and Downfall...................155
CHAPTER 19 — Atomic Warfare................................161
CHAPTER 20 — Japan's Response..............................171
CHAPTER 21 — The Road to Peace...........................173
CHAPTER 22 — The Palace Coup..............................181
CHAPTER 23 — The Non-Surrender.........................187
CHAPTER 24 — The Aftermath and the Math.........191

CHAPTER 25 — Japanese Technology..................199
CHAPTER 26 — Marcus Island..............................211
CHAPTER 27 — The Golden Lily..........................215
CHAPTER 28 — MacArthur....................................221
CHAPTER 29 — The Surrender..............................229
CHAPTER 30 — The Tribunals...............................235
CHAPTER 31 — The Roswell Incident...................239
CHAPTER 32 — The Coverup.................................255
CHAPTER 33 — 1947...261
CHAPTER 34 — Bargain with the Devil................265
CHAPTER 35 — International Recognition...........273
CHAPTER 36 — Conclusion....................................283
EPILOGUE — by Peter Moon................................287
APPENDIX A — The Dread Zeppelin Legacy......293
APPENDIX B — The San Antonio Crash.............299
APPENDIX C — Attachment by Wouter Hobe....303

BY PETER MOON

FOREWORD

It is normal and routine for any reasonably well educated person in today's world to believe that the Japanese lost World War II, unconditionally surrendering to the Allied Powers on September 2, 1945 in a signed ceremony aboard the *U.S.S. Missouri*. While there was indeed a ceremony on that date aboard the *Missouri*, nothing could be further from the truth about an unconditional surrender or a genuine surrender of any kind. The idea of an unconditional surrender or even a Japanese surrender is a myth that was very cleverly created by the Office of War Information but one which has also been pervasively carried forth by its successors. The myth lives on in direct proportion to the truth that has perished.

The proof of such a claim can be found by a routine search of history. For example, most people do not realize that President Harry Truman did not announce the cessation of hostilities until the last day of 1946, well over a year after the Japanese supposedly signed an unconditional surrender. Further, a formal treaty of peace was not signed with the Japanese until September 8, 1951, but it did not go into effect until April 28th, 1952 in celebration of the Japanese Emperor's birthday. This is all historical record.

Further, when a nation surrenders unconditionally, it is not in a position to negotiate a peace treaty of any kind. It simply submits to the will of the conquerors. Further still, Japan paid no war reparations of any kind to the United States. In fact, the United States was obligated to clean up the atomic rubble and radiation it had unleashed on Japan, many American soldiers dying as a result.

If none of the above convinces you, consider that Japan became a member of the United Nations Security Council, the very organization which had been sprung into being, as an organization of war, in order to defeat the Empire of Japan. When you defeat an enemy, you do not allow them into your board room, and this is exactly what all of the Allied powers did. As a result of their own subjugation to the United Nations, per the U.N. charter, the United States is no longer allowed to declare war. This is also why the Department of War was abolished and replaced by the Department of Defense.

Like most people, I would have never have even realized the existence of those facts, their implications, or even the obvious contradictions

BY PETER MOON

concerning World War II propaganda if it were not for Douglas Dietrich, a research librarian for the Defense Department who was assigned to burn government documents that were top secret and beyond. None of the above history about peace with the Japanese, however, has anything to do with classified or top secret information. As was said, you can find it in ordinary history. What you will learn in this book are not only more details regarding the above, but a slew of information that presents an entirely different paradigm that the world has been ignorant of and that the academic world obfuscates. As all of the media, including the academic world, is embedded so as to serve as propaganda agents on behalf of the military, your chances of learning this information would have been virtually impossible if not for Douglas Dietrich.

On a personal note, you might find it noteworthy that, as a young boy, my father would frequently take me fishing and we would pass close by the huge ships that were docked in the port of Los Angeles. Amongst these were many Japanese vessels that featured the Japanese flag, a red circle on a white background. My father, who was a first hand witness to the Pearl Harbor attack and did what very little he could to fight back, explained to me the difference between the Japanese flag and its war time flag, a red circle with red rays extending in all directions.

It was not unusual to see a Japanese war ship, all of which featured the regular Japanese flag. One day, however, I saw something that was surprising. It was a war ship that featured a Japanese war flag. I pointed it out to my father, and he was also surprised. He said he would find out about it, but nothing ever came of his efforts. Upon the compilation of this book, I did a search and came up with the following photographs of Japanese vessels with actual war flags on them in Los Angeles Harbor in 1959. While I cannot turn up a photo of the battleship I saw with a war flag, this evidence tells me that my memory was quite correct. Why, however, were Japanese war ships in Los Angeles Harbor? My best guess is that they were sending a strong message which resulted in the *Treaty of Mutual Cooperation and Security Between the United States and Japan* that was signed in 1960. This limited the powers of the United States in the Far East.

The following pictures are from the University of Southern California library, sleeve number 12608=009, originally from the *Los Angeles Examiner* (Legacy Record ID examiner-m19077) in the sub-collection "*Los Angeles Examiner* Negatives Collection, 1950-1961". The photos are accompanied by the following caption:

FOREWORD

2 images. Japanese ships in Los Angeles Harbor, 28 July 1959. Rear Admiral Jiro Akahori (Commander of Japanese Training Squadron); Commander R. Ripley; Yukio Hasumi (Japanese Consul General); H.M. Sato; three ships of Training Squadron.; Supplementary material reads: "Gershon; city desk; illus; Jap ships. Shots of Rear Admiral Jiro Akahori, Commander of Japanese Training Squadron salutes and Commander O.F. Ripley, representing Rear Admiral Walter H. Price, Commander, U.S. Naval Station, with Japanese Consul General Yukio Hasumi and H.M. Sato, president Japanese Chamber of Commerce and head of Japanese colony welcoming committee".

BY PETER MOON

There is also another matter with regard to my own published works. After publishing *The Montauk Project* in 1992, it was only a few months before I was contacted by an agent for Gakken, the largest publisher in Japan, in order to arrange for publication of the book in the Japanese language. A Japanese television documentary was produced almost immediately. Since that time, I have always found it a curiosity that when it comes to international rights, my various works have only been published in countries that once belonged to and/or were closely tied to the Axis powers. These include Japan, Germany, China, Russia, Romania, Bulgaria and Spain, the latter officially being a neutral country which was, for all practical purposes, a close Nazi ally.

It was only after encountering Douglas Dietrich and studying much of his prolific information that I fully understood the causes of the aforesaid unusual circumstances and events in my life.

Some of you might also remember the 1980s when strong anti-Japanese sentiment was aroused in the United States as a result of so many Japanese, whose economy was booming, visiting as tourists and also buying up huge tracts of land, including Rockefeller Center in New York City. It was as if the roles were reversed, at least as far as war propaganda goes. People were boldly asking the question: "If we had won the war, then why were the Japanese the benefactors?"

As I have already stated, this book will answer this and many more questions, including questions you never thought to ask. Besides this, I would also like to make another point. Besides new insights into the actual dynamics of World War II and what was behind the Roswell Incident, this book will also introduce you to an ancient culture that is over 2,600 years old. Ever since I was a young boy growing up in Southern California, I routinely heard how uncultured Americans generally are. Part of the problem is that, since the inception of the United States, we have had a relatively uniform culture throughout much of the United States. While there are pronounced differences between the northern, southern, western and eastern parts of the country, there are many commonalities, language and Christianity being the most prevalent.

In Europe, a much smaller area than the U.S., there are many different languages and customs. As these different regions are so close and well-traveled, it is far easier for Europeans to become "cultured" and well-informed about their neighboring regions. The countries in Asia, however, are in complete contrast to all of this. Its culture and history are not well understood and are even distorted beyond belief. Whatever you might think about Asian culture as a Westerner, and particularly that of the Japanese, it is a far cry from the Asian mind. One among many of these aspects concerns Buddhism, an ever pervasive

FOREWORD

staple of Japanese culture. Writer Jeremy Mohler has very appropriately described one's Buddha nature as being representative of one's very humanity and our ability to be courageous and open to what we do not know. What Douglas Dietrich has offered us in this work is not only a very deep look into the Asian perspective but into the heart of humanity itself and the warrior mind-set that has perplexed it. In reading this book, you are advised to be courageous about what you do not know. That is the only way you can learn.

Peter Moon
Long Island, New York
April 19, 2021

BY DOUGLAS DIETRICH

INTRODUCTION

In order that you understand the unique perspective and circumstances which have exposed me to the information I am to share in this book, it is advisable that I first give you a brief biographical account of myself. I was born in Taipei, the capital city of the National Republic of China, on the island of Taiwan, the territorial status of which is disputed to this day. Although it is not known to most Westerners, Japanese is the common language on this island; and Taiwan, formerly known as Formosa, was a Japanese colony from 1895 through the end of World War II.

My mother, Dianna Sūjĭn-Lín ("Bright Woods" or "Forest 'o' Brocade") Dietrich (ne Takabayashi Hideko or "Splendid Child") was both Japanese and Chinese by ethnicity and her knowledge of multiple languages, in addition to her descent from a royal Chinese bloodline, prompted Emperor Hirohito to appoint her as a diplomatic interpreter for the imperial Japanese envoy that personally mediated with Adolph Hitler. While my mother's first-hand experiences gave me an uncommon perspective on the politics between the Allied and Axis powers, my father's life experience was equally unique.

George Joseph Henry Dietrich had joined the Navy at the age of sixteen during a time when recruits were hard to come by and such violations of law were routinely overlooked. He served in the gun boat patrols on Chinese rivers during the 1930s. America's own military position during this period has been dramatized in the movie *The Sand Pebbles* with Steve McQueen, a film which gives a fair representation of what actually went on as American sailors were fighting the Nationalist Chinese to the death. Having also served in the Korean and Vietnam wars, my father had extensive experience in Asia and knew the history of the region in a way that most soldiers do not.

During my teenage years, I was attending John O'Connell High School in San Francisco as a military dependent. At this time, I had come to know a lady who was working at John O'Connell High. Her name was Lee Ann Prifti,[*] and she was aware that my status as a military dependent gave me a military I.D. card with access to San Francisco's

───────────────
[*] Prifti had previously worked at city hall and was very close to the mayor of San Francisco, George Mosconi. She was present when Mosconi was shot in 1978, she herself being pushed to the ground by the assassin, Dan White.

military base at the Presidio. When she told me of a job opening for a Librarian's Aide at the Defense Department's Research Library located on the base, I applied for the job and was accepted.

While serving as a Librarian's Aide at the Presidio, I graduated from John O'Connell early and went to the City College of San Francisco where I also graduated early, eventually accumulating enough credits for a Bachelor of Science degree. Thanks to my education and aptitude for fast learning, I was promoted to the status of Military Reference Technician. Within two years, I became a fully fledged Department of Defense Research Librarian by the age of eighteen. What is most significant to this narrative is that, throughout that entire time, I was exposed to increasing levels of top secret and highly classified documents.

The Department of Defense is not a military organization but rather a civilian institution, and in the case of the Presidio, the Department of Defense Library happened to be located on this particular military base. My status therefore was that I was a civilian working on a military base.

One of the absurd conditions of working for the Government is that you can never be fired, and the employees took full advantage of this. It was typical for them to come to work drunk; and in the case of one particular individual, it proved to be an absolute disaster when he came to work in such a state, absolutely saturated with alcohol. This man went to work in the basement where it was his job to destroy sensitive documents by burning them in an incinerator which was actually a

OVERVIEW OF THE PRESIDIO IN SAN FRANCISCO

INTRODUCTION

modified crematorium. In his inebriated state, he clumsily fell into the incinerator where ninety per cent of his body was burned. That was not only the end of him at the Presidio, but this horrific incident induced a loathing of the area that became so extreme that no one wanted to go near the incinerator. Consequently, the materials to be destroyed continued to accumulate for a period of up to ten years.

It was into this legacy that I was handed my first job as a Defense Department Research Librarian. Ordered to destroy these documents and knowing very well these were well above my own personal security level, I protested this assignment vehemently. My supervisor, whose motive for assigning me to this task was due to his own laziness and reluctance to do this dirty work himself, told me in very certain terms that, if I refused the assignment, he would go to the military police and tell them I had attempted to abscond with these materials, the result being that I would be put in jail for near life. It was therefore, by reason of this and under protest, that I began the process of destroying these highly classified documents.

The documents I was ordered to destroy were often in old locker boxes, wooden with metal edges, and were loaded with official records from the Spanish American War all the way through World War II and Vietnam. It also included a host of other documents, and these included the so-called Roswell Incident, the mysterious crash in 1947 which happened over Roswell, New Mexico, an event which has been popularly credited as being the result of a flying saucer having suffered a crash landing. In the case of the Roswell documents, I was not ordered to destroy them but to collate, verify and review them so that they could be presented to the commandant of the Presidio military base himself.

Amongst my many other responsibilities, I destroyed classified documents for over a decade, but it was all under protest. I did, however, take the opportunity to make mental notes with my memory and made copious handwritten notes when I returned home each night. These included sketches that were turned into elaborate illustrations of the giant dirigibles that the Japanese used before and during World War II and have remained a great secret ever since. As you will read in this book, they were an integral part of the Roswell Incident.

What I have said here will give you a background as to how I accessed the data behind the narrative of what is presented in this book. All of it was supplemented by what I learned from my father and mother as well as my general experience in and out of the Presidio military base.

I would also add that, due to the nature of the information I release on a regular basis on the internet, which extends far beyond what is presented herein, I have been repeatedly monitored by interested parties who perceive themselves to be threatened by what I expose. This includes being trolled extensively on the internet and having the telephone/cable box outside by home routinely doctored with different technology. These claims are well documented. If what I have to offer can just be laughed off as speculative hearsay, there would be no motive for the extent of monitoring and gang-stalking that is directed at me. What I offer to you here can be validated by doing your own research.

As you read this book, it is important to keep in mind that America and virtually all of the world have been subjected to a continuous and repeated propaganda campaign concerning the facts and actual history of World War II, all of it beginning with the Office of War Information at the onset of the war and extending throughout Hollywood and the media, all of which is embedded so as to serve the interests of military propaganda.

I will begin this book by making an opening statement of what I intend to convey to the reader, much like the manner in which an attorney might present an initial argument in a court of law. In the end, you yourself and the other readers will serve as the collective jury, all of which will determine the relative impact this book will have on the world at large.

BY DOUGLAS DIETRICH

OPENING STATEMENT

What I intend to present in this book is a deconstruction of two major myths that have been perpetrated upon the world and particularly America, both of which have had an extremely negative impact with regard to understanding history as it really was and current events as they actually are.

Beginning with Elmer Davis and the Office of War Information (which was eventually absorbed into the State Department and the CIA), an extensive false mythology has been created surrounding the actual circumstances and military facts of World War II. Through control of media outlets, including Hollywood, a myth has been created and successfully indoctrinated into the minds and routine thinking patterns of the population, convincing the public of the following:

Myth #1: The Empire of Japan was an aggressor nation who attacked Pearl Harbor without provocation nor just cause; and further, as a direct result of atomic bombs being dropped on Hiroshima and Nagasaki, Imperial Japan surrendered unconditionally, it being announced by Emperor Hirohito on August 15 and formally signed on September 2, 1945 aboard the *U.S.S. Missouri*, thus bringing the hostilities of World War II to a close.

Myth #2: Based upon a July 8, 1947 press release from Roswell Army Air Field (RAAF) public information officer Walter Haut, the seeds of a myth were created when a story was circulated in the *Roswell Daily Record* that personnel from the field's 509th Operations Group had recovered a "flying disc" which had crashed on a ranch near Roswell. This included a report that an Intelligence Officer of the 509th Bomb Group, stationed at Roswell AAF, Major Jesse A. Marcel, had recovered a "flying disc" from the range lands of an unidentified rancher in the vicinity of Roswell and that the disc had been "flown to higher headquarters". That same story also reported that a Roswell couple claimed to have seen a large unidentified object fly by their home on July 2, 1947.

BY DOUGLAS DIETRICH

As a continuance of this myth and adding complexity to it, the July 9 edition of the *Roswell Daily Record* featured two headlines: "Ramey Empties Roswell Saucer" and "Harassed Rancher Who Located 'Saucer' Sorry He Told About It".

The first headline referred to Brigadier General Roger Ramey who stated that, upon examination of the debris recovered by Marcel, it was determined to be a weather balloon. The wreckage was described as a "..bundle of tinfoil, broken wood beams, and rubber remnants of a balloon."

The additional story of the "harassed rancher" identified him as W.W. Brazel of Lincoln County, New Mexico who claimed that he and his son had found the material on June 14th when they "came upon a large area of bright wreckage made up of rubber strips, tinfoil, a rather tough paper, and sticks." He picked up some of the debris on July 4 and "..the next day he first heard about the flying discs and wondered if what he had found might have been the remnants of one of these." Brazel subsequently went to Roswell on July 7 and contacted the Sheriff who apparently notified Major Marcel. Marcel and "a man in plain clothes" then accompanied Brazel home to pick up the rest of the pieces.

Over the years, reports by additional witnesses and disingenuous but official military accounts of it being a weather balloon gave rise to the prospect that the military is deliberately hiding the truth. Further reports, based upon alleged eye-witness accounts, give rise to stories that a mysterious retractable "metal" had been recovered at the crash site and that the craft featured unintelligible hieroglyphics. Further accounts speak of yellowish orange or gray hairless bodies with three fingers being recovered near the crash which were placed in child-sized coffins which had been ordered by the local mortuary.

Eventually, the myth of aliens crashing at Roswell became institutionalized with the International UFO Museum and Research Center being erected at Roswell, all followed by new public clamor as a book entitled *The Day After Roswell* is published, written by Phillip Corso and Bill Birnes. Corso, who was a Special Assistant to the head of Army Research and Development, Lt. General Arthur Trudeau, and was in charge of the Foreign Technology Desk, claims that modern technology has been, in part, the result of reverse engineering technology that was discovered in the Roswell craft.

DECONSTRUCTION

It is the intention of this book to deconstruct both of these myths which are intricately woven together by the nature of the Japanese

OPENING STATEMENT

attack on the United States and the actual events and circumstances that occurred before, during and after the so-called Roswell Incident. The war did not begin with Pearl Harbor. While I will go into greater detail, it will be established that, in addition to initiating and conducting a long time policy of aggression against the Empire of Japan, the United States deliberately triggered hostility and an actual state of war against the Japanese as early as 1937 in order to protect the opium trade in China. As opium was undermining the Chinese economy and the Japanese were trying to establish economic relations with China, the Japanese had interfered with the American sponsored drug trade. American response was to have their soldiers dress as Chinese coolies in blackened uniforms in order to instigate a guerrilla ambush against the Japanese during peace talks they were then having with the Chinese.

Believing they were betrayed and that the ambush was generated by the Chinese themselves, the Japanese responded on December 9, 1937 with what became known as the Rape of Nanjing which resulted in hundreds of thousands of deaths. Through interrogation of Chinese prisoners, it took the Japanese 72 hours to figure out that the instigators were Caucasians and not Chinese. This resulted in the Japanese bombing the American ship *Panay* and Her Majesty's ship the *Lady Bird* on December 12, 1937. This began an actual state of war between the Americans and the Japanese and was referred to as such, both colloquially and in the newspapers. The Office of War Information, however, suppressed the actual news and the Congress did not declare war because the facts, if revealed, would badly embarrass the Americans to no end. The Pearl Harbor attack was a continuance of the already ongoing hostilities between the two countries, but it gave Roosevelt an ostensibly justifiable reason to declare war, all the time obfuscating the original cause.

The very successful Pearl Harbor attack was followed up two months later when Japanese super-dirigibles invaded the air space of Los Angeles on February 25, 1942, triggering a massive panic in both the population and the military. The Japanese strategy was not to bomb Los Angeles but to frighten the American military brass into believing they were going to infect the American population with biocidal weapons of mass destruction. The Army responded immediately with massive artillery fire, downing some of the Japanese air craft, and fast action was taken to truncate any reports of a Japanese attack. The Secretary of the Navy, Frank Knox, attributed the American response to a case of war nerves with no actual attack having taken place nor any foreign or otherwise unidentified aircraft having been present.

BY DOUGLAS DIETRICH

As the President and the military brass had good reason to believe that Emperor Hirohito and the Japanese were experimenting with yellow fever as a biocidal weapon, they incorrectly assumed that he was releasing such on the population of Los Angeles. Accordingly, they created an instant and erroneous vaccine which killed 50,000 of their own men with hepatitis and infected another 350,000 who had to be mustered out of the service because they were unfit for combat, eventually dying years later.

The super-dirigibles, which had been originally developed to access the otherwise inaccessible hill country of China, consisted of one huge hydrogen balloon surrounded by nine smaller balloons. With a lift capacity of 250 tons, these super dirigibles were capable of carrying small planes and mini tanks as well as troops. The typical operators were small in stature, often not greater than four feet in height. These were the Yakuza whose tradition includes the severing of the pinky finger in order to atone for their sins.

During the atomic bombing of Hiroshima and Nagasaki, no military targets were hit and the Japanese military was completely unharmed as their forces were all based on the Asian mainland. It took only a matter of days for the Japanese to respond by sending super-dirigibles to cross the Pacific by reason of the Kuroshio or Japanese Current. Three of these, all piloted by the four feet tall Yakuza, were sent to Tonopah Army Air Field in Nevada, two of them landing in the area which is now famously known as Area 51, one of them having crashed en route. The Yakuza were treated as prisoners of war, and despite the protracted peace negotiations with the Japanese, the presence and status of these soldiers was never officially nor otherwise acknowledged by the United States, a war crime which is, in fact, still extant.

These super-dirigibles, which included actual biocidal weapons of mass destruction, were sent by Hirohito as a threat, in disambigution from a strike, and this was done to induce the Americans into negotiations for peace, a request that was not acceded to and realized until September 8, 1951, when a treaty of peace, known as the *Treaty of San Francisco*, was finally signed between Japan and the United States. The treaty itself went into effect on Emperor Hirohito's birthday, April 28, 1952. It was a day of celebration for the Empire of Japan because the economic doors of the West were now wide-open to them, a prospect which was denied to them prior to the war, the United States then having sought a "colonialization under threat of strangulation" policy against the Japanese.

While peace negotiations ensued after the so-called surrender aboard the *USS Missouri*, President Truman did not declare an official end to

OPENING STATEMENT

hostilities until the last day of 1946. In the meantime, the Americans sought to operate the super-dirigibles for their own purposes, but they were too physically large to even occupy the flight control centers. Accordingly, they tortured the Yakuza and forced them to experiment with these craft on behalf of an American agenda. During flight tests over the Four Corners region; and after winds had blown the craft towards Roswell, the Yakuza self-immolated rather than be subjected to continued imprisonment and torture at the hands of the Americans.

The American brass were guilty of horrible war crimes by reason of keeping these soldiers as prisoners of war and torturing them, but this was only the tip of the iceberg. If it were to become known that the Japanese had such sophisticated aero-technology, to say nothing of the genocidal weapons, it would lead to a domino-effect with thousands of questions, the end result being a national and international embarrassment that was beyond belief.

In order to avoid complete embarrassment, the Government sought a strict policy of containment and demonstrated it with an iron first, never hesitating to threaten the lives of witnesses and their families. Their first effort at damage control was to facilitate a press release announcing that this crashed aircraft, which was in fact a Japanese super-dirigible, was a flying saucer. Cleverly retracting their original release a day later, the Army left a distinct impression they were lying, stating it was only a weather balloon. Ever since, this lie has been steadily built upon, leading to the creation of an entire "alien-inspired subculture" to the point where this myth has been a closely cherished staple of the UFO community and the American public in general. Roswell is now, in-effect, a shrine to this U.S. Army myth.

While this is only a general outline, what will be presented here is a full load of information which will not only back up the case but reveal a far more sinister set of circumstances than the worst alien horror story you might imagine. In the end, the truth will be undeniable to anyone who takes the time to absorb the facts and evidence presented in this book.

BY DOUGLAS DIETRICH

CHAPTER ONE

JAPAN - AN ISOLATIONIST EMPIRE

To understand the antecedents of World War II and what precipitated it, it is very important to grasp at least a bit of the fundamental dynamics of the culture of the Empire of Japan and the Japanese people, a civilization that goes back well over two thousand and six hundred years prior to their preemptive strategic surgical strike on Pearl Harbor. Japan possesses a vast history compared to the mere one hundred and sixty-five years that the United States of America had been in existence prior to the declaration of war.

An island empire, Japan has thrived on being an isolationist culture since its inception, but it has nevertheless suffered continuous assaults and invasions in attempts to take its resources and subjugate its population. One of the most prominent instances in this regard was in the 13th Century when the Mongol Empire, then under the rulership of Kublai Khan, instigated several skirmishes, all of which the Japanese successfully defended. Eventually, a Pan-Asian attack by thousands of Chinese-built and Korean-crewed ships transporting Mongol Marines was vanquished when a great typhoon arrived, contributing to the sinking of over 4,400 ships (equaling in contemporary composition the over 4,000 ships deployed by the Allies on D-Day) with over 70,000 men drowned in what was the greatest loss of life at sea in recorded history. This failure resulted in the Mongols withdrawing for seven years, but it invoked the supreme wrath of the great Kublai Khan, and the Mongols turned their full attention on Japan with an entire ministry being set up whose sole purpose was to conquer the Japanese.

As the Mongol-Chinese Empire was huge, the proposition of such a war was most decidedly a lost cause for Japan with no hope of victory by normal military means. Due to these desperate circumstances, Emperor Kameyama-Jokō personally prayed at the Grand Shrine of Ise, asking for intervention on Japan's behalf by the great Shinto deity, Amaterasu ("He who is said to shine in heaven"). This entreaty by the Emperor is historically recorded as having been made on 15 August 1281.*

---

*664 years later, Emperor Hirohito would use this exact same date, 15 August 1945, to accept an overture of peace by the Allies. Specifically, he accepted the *Potsdam Declaration* on the condition that it would not compromise the sovereignty of the Emperor himself. In other words, he remained completely in charge of his country. The nuanced political implications of all this will be explained throughout the book.

The historicity of this event has been compounded and intermingled with legend and myth, but it is undisputed that the Khan's massive fleet of 3,900 ships and 140,000 soldiers, who were traveling only a very short distance from the Asian mainland to Japan, were completely wiped out as the result of what are known as the Kamikaze or "Divine Winds", also identified as typhoons. This prayer of the Emperor has ever since been a source of national pride in recognition that he successfully called down the Kamikaze as Divine Winds of protection to save the Japanese Empire and all of its subjects.

An attack on the sovereignty of the Empire was experienced again in 1853 when Commodore Matthew Perry invaded Japan with his steam engine navy and aggressively forced Japan, by reason of his superior armaments, to open its markets to the outside world. Perry's mission was backed by American industrialists, including railroad barons, who wanted a rail line from Japan to Korea and up into Siberia and across the entire Asian continent. The Japanese realized that if they did not participate in the industrial revolution, they would eventually be completely consumed by Western culture.

COMMODORE MATTHEW PERRY

AN ISOLATIONIST EMPIRE

The Japanese had already seen what the British had done in China by instigating the opium trade, forcing each and every Chinese citizen to buy a required allotment of opium per month. The ever competitive Americans were making inroads into Asia to compete with the British because they wanted their own foothold and thus invaded the Philippines at the turn of the century. This not only gave the United States access to the rich oil fields of Indonesia, it also restricted Japan's access to such.

The sentiments of American aggression were echoed to an even greater extreme when a heralded American literary figure by the name of Jack London published a short story in *Collier's* magazine entitled, *The Unparalleled Invasion*, a tale which created and fostered the idea of the "Yellow Peril", a term which refers to the prospect that the Chinese or "Yellow Race" is menacing both racially and militarily and are a threat to the "civilized world" of the West. What was especially offensive and inhumane in London's story was that it advocated the complete elimination of the Chinese as a people.*

Although the origin of the term "Yellow Peril" is frequently attributed to Kaiser Wilhelm II of Germany, its earliest known appearance in print was from the Hungarian General Turr in several US newspapers at the time. The following is from the Ohio paper *The Sandusky Register* of June 1895:

> "The 'yellow peril' is more threatening than ever. Japan has made in a few years as much progress as other nations have made in centuries."

Besides demonstrating the fear of the West, this quote demonstrates how technologically advanced the Japanese were as a civilization. It

---

*Jack London, a native and favorite son of California, was very much influenced by the culture of his time and place. You will better understand his mentality by reason of the fact that in the mid-1800s, the state of California, being concerned about Asian immigration, produced more legislation against Chinese immigrants than against African Americans. When California became a state, Mormons served on the first legislative body, and the first law they created was that no native, black or Chinese could serve as a witness against a white person.

With the Union victory of the Civil War prohibiting slavery from expanding into the West, the door was open to immigrant labor for particularly dangerous and low paying jobs. Jim Crow laws were created to chip away at the civil rights and liberties of Asian Americans, immigrants and anyone who was not white. From the 1850s until the 1960s, Jim Crow legislation attacked their rights to due process (the right to testify in court), which had the same effect as legalizing the murder of non-whites, since witnesses to those crimes could not legally testify in court. Jim Crow laws denied people of color their civil rights to hold office, vote, marry a Caucasian (miscegenation), receive an education, employment, consume alcohol, or choose where they live. Jim Crow laws also forced segregation.

JACK LONDON

was the Japanese, following Commodore Perry's invasion, who issued an edict to get rid of the barbarians from the West. Much has been misconstrued about the technological prowess of the Japanese, all leading to supreme ignorance and a denial which has had a very adverse effect on the American public as well as their government. In this book, you will learn that an ancient civilization has certain built-in advantages over their adversaries.

As *Collier's* was an international magazine that was translated across the world, Jack London's *Unparalleled Invasion* found its way into the hands of a young Hirohito who was completely alarmed by what he read. Based upon the history of Western behavior and the open and tacit threat to annihilate the Chinese, the Emperor to be knew that such an imperative could easily apply to his people as well. After Japan's defeat of China in 1895, the term "Yellow Peril" was also applied to the Japanese.

With regard to the history of World War II, no character has been more misunderstood than Emperor Hirohito and his actual role in the war, a subject which was strategically neglected as the war ended and has never been addressed since. Although the Emperor was THE major player in the war, his role is ignored, but you will learn about it in this book.

AN ISOLATIONIST EMPIRE

**EXAMPLE OF RACIST LITERATURE OF
THE TIMES THAT DEMONIZED ASIANS**

CHAPTER TWO

EMPEROR HIROHITO

Perhaps the most obvious and easy to dispel obfuscation that the media engineers have saturated the minds of Americans with is the complete fiction that Hirohito, the Emperor of Japan, was neither a war criminal nor was a significant player with regard to waging war in the Pacific or in the United States itself. As you will learn herein, Emperor Hirohito was not only engaged at the very beginning, he was directly responsible for engineering the entire circumstances by which the war could actually come to an end. Only one indication of this is that he devised for the peace treaty between the Japanese and the Allies, including the United States, to go into effect on his very birthday.

As a young child, Hirohito was sequestered from his parents and tutored under General Nogi Maresuke, a great warrior who had brought the second Russo-Japanese war to a successful conclusion. Nogi was a samurai in the truest and most traditional sense of the word. To understand the psychology of Hirohito, it must be understood that his father deliberately chose for him to be trained in the tradition of the samurai. This meant that it was Nogi's task to teach the young prince with beatings and severe discipline as this was deemed absolutely necessary to make an effective warrior. Part of this tradition is that fathers, whether they were noblemen or not, did not have it in their heart to beat and discipline their own children in such a manner so as facilitate them to become formidable warriors. Hirohito's siblings were not given this distinct "samurai honor" of being so trained.

Under these circumstances, Nogi did his job so well that, as their relationship developed, his royal pupil came to unnerve him. This was pointedly demonstrated when the young prince read Jack London's *The Unparalleled Invasion*, the horrific story of genocide that prompted the young Hirohito to take stock of his budding enemies in the West. *The Unparalleled Invasion* not only advocated genocide based upon democratic principles, it called for the extermination of every single Chinese man, woman and child.

Upon reading this appalling story, Hirohito turned to his mentor, General Nogi, and said the following, "Your samurai swords, your weapons of guns, planes and tanks are useless. The Americans will kill us with germs. If they're willing to do it to the Chinese, they will do it to us."

YOUNG HIROHITO

As a result of this, Hirohito dedicated his life to marine biology and became an internationally recognized marine biologist, even earning a Ph.D. in the subject. I learned about this as a Department of Defense Research Librarian when I was ordered to destroy thousands of documents and photos, and these included Hirohito in a lab smock working with a microscope. Besides earning a Ph.D., he even discovered a new species of animal. In the days before we had satellite imagery and the new technology we have today, discovering a new species of animal made your career. Although this is not broadly known in the West and even suppressed, he was internationally renown as a marine biologist, and he used this knowledge to start developing biological weapons that would kill the Americans before they could kill his people. The photos I

NOGI MARESUKE

saw of Emperor Hirohito working on weapons of mass destruction were the only photos I have seen with Hirohito smiling. No one, however, ever hears about this nor sees him with the instruments of his chosen career specialty. As a commercial illustrator trained at John O'Connell Vocational Institute, I sketched copies of the photographs I was to burn.

Hirohito's passion for biological warfare in his early life distressed his mentor very much. Both frightened and upset by the young Emperor to be, Hirohito was someone the likes of which he had never seen. To Nogi, Hirohito was neither a samurai nor a warrior and the teacher viewed his pupil as a psychotic terrorist. It was this conclusion and his other convictions about his servitude to the Empire which ran so deep in his soul that prompted General Nogi to kill himself. After this, in his personal diaries, Hirohito wrote, "General Nogi was a very brave soldier for the old kind of war, but he is a coward for the new kind of war."

A YOUNG EMPEROR HIROHITO, MARINE BIOLOGIST

    Although Nogi did not recognize the Emperor as a samurai, Hirohito's dedication to his empire was very much in this tradition, and this can be gleaned from an ancient text entitled *Hagakure — The Book of the Samurai*. Literally translated as "Hidden by the Leaves", *Hagakure* is a compilation of commentaries from 1709-1716 by Yamamoto Tsunetomo concerning the problem of maintaining a warrior class in the absence of war, a situation which had caused Japan to become vulnerable to outside invasion, such as in the case with Commodore Perry.

    Hagakure became a staple of Japanese philosophy in the Pacific War. Two different copies of the book below reveal that the emblem for this work depicts three cranes with the wings interlocking. This is representative of the Sacred Crane ("Showa" in Japanese, a name that the Emperor assumed upon taking the throne). The profundity of this symbol and the impact that Hirohito's biological weapons program had on the world was that this symbol is now universally used as a warning for materials, cargo, or waste that contain biohazards.

    Hirohito's vanguard for his biological weapons strategy was known as the Sacred Crane Task Force, and its most infamous and notorious branch for such bio-warfare was Unit 731, set up directly under the orders of the Emperor. Featuring a special armband with the image of a black sun, this symbol represented the black budget that was directly under the imperial budget. While there are entire books written on Unit 731, none of them address the hidden aspects which you will read about in this work. Before we address any of that, however, it is first necessary to get a bigger picture of Hirohito's impetus for fervently embracing biological warfare of the most sinister kind.

Although it is not broadly recognized and was most certainly suppressed by the United States Government, World War I itself was ended as a direct result of biological warfare waged by the United States. I will address this in the next chapter and how it influenced World War II and its key players, all of it ultimately and circuitously leading to a crash near the Roswell Army Air Field.

ABOVE IS THE INTERNATIONAL SYMBOL FOR BIOHAZARDS WHICH WAS ADOPTED FROM "HAGAKURE". ON THE FOLLOWING PAGE YOU CAN CLEARLY SEE HOW THE SYMBOL OF THE EMPEROR'S "SACRED CRANE" SYMBOLISM INSPIRED THE ABOVE SYMBOL.

THE ROSWELL DECEPTION

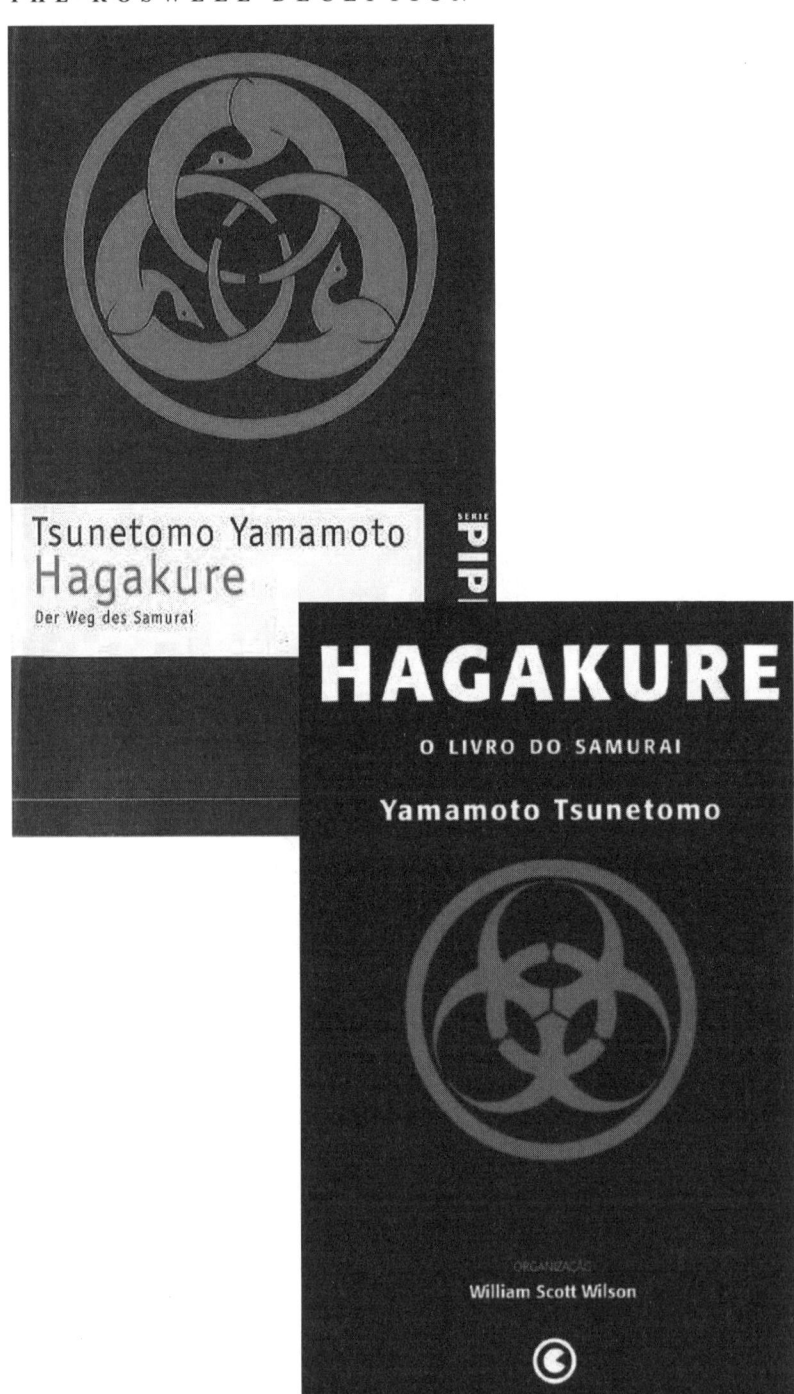

CHAPTER THREE

THE AMERICAN ARMY AND NAVY FLU

As open conflict wound down in the European theater of war in 1945, the Allies consternated over the official edict and surrender terms they would inflict upon occupied Germany. The reason for this is that it was widely believed, and most certainly asserted by the Germans themselves, that the reason Germany was able to resurrect itself as a war power after World War I was due to the fact that the *Treaty of Versailles*, which put the Germans at a severe economic disadvantage, was signed by the civilian government and not the military. In other words, they defeated the former but not the latter. Hence, the Allies did not want to repeat the same mistake in 1945. Instead, however, the Allies did the reverse. They got the military to accept a surrender, under the authority of Admiral Karl Dönitz, but they never officially nor legally dealt with the duly elected civil government of the German people: the National Socialists.

Consequently, the Allied Powers and the United States of America are still technically and legally in a state of war with the National Socialists of Germany, many of whom moved to the Southern Hemisphere where they maintain an occulted but active presence to this day. You can still find very real evidence of this fact in your telephone bill as every month there is a phone tax that is still in effect. We have also remained on Daylight Savings Time, a law signed and put into effect by Woodrow Wilson during World War I and later adopted by Franklin Roosevelt during World War II.

This history is important because it gives context as to how World War I set up the circumstances of how World War II played out. Up to now, I have only scratched the surface of what I am about to expose as a result of reading classified documents, so much of which dealt with biological warfare. To understand the history of biological warfare in America, we have to go back to well before the Revolutionary War and to the time of the Salem Witch Trials.

During the days of America's theocracy, going back to shortly after the landing at Plymouth Rock, Cotton Mather was an infamous and despicable character who is mostly known for his active role in promoting the prosecution of witches during the Salem Witch Trials. Advocating the extinction of all Native Indians as the spawn of satan, Mather lived during the time of a massive outbreak of small pox, a disease which ravaged the Native American population and was widely used as a weapon of mass destruction against them. Many of you are familiar with the history of gifting the Indians with blankets to infect them with small pox.

COTTON MATHER

Mather, who was a puritanical fire and brimstone preacher, was the leader of a congregation who gifted him a slave by the name of Onesimus. Onesiumus explained to Mather how he had been inoculated as a child in Africa, and this fascinated Mather to no end, whereupon he became a leading advocate of vaccination which, at that time, was referred to as variolation. Although Mather faced an uphill battle against the anti-vaxers of his day and many deaths were suffered, the procedure became a staple of American medicine, but it also had a sinister side to it by opening up the door to biological warfare. Although variolation or vaccination provided a degree of immunity, small pox could also be used to contaminate blankets and facilitate genocide against the Natives.

Small pox reared its head again in the days of what was known as Transylvania Colony, a huge tract of land that took up most of what is today known as Kentucky. Many of the founding fathers of America had purchased this land under the auspices of the Transylvania Company, an act which in itself served as the major catalyst that caused the actual insurgency culminating in the Declaration of Independence in 1776.

This was the world's first revolution of the modern age that was based upon a lot of conservative plantation owners who owned slaves and wanted more land. King George III had already drawn a proverbial line in the sand at the border of where Transylvania Colony began, at the Cumberland Gap, and said "this far and no further". It is called the Royal Proclamation and stated that the American colonies had to remain behind the Appalachian mountains and were not to go further west in order that the Americans Indians could live on in peace. The American founding fathers said no and launched a war that is known as the American Revolutionary War of Independence. The primary plan was to get the land that they wanted and to exterminate the Native population, George Washington being a major proponent of this agenda.

One of the biggest problems for those plantation owners was running their empire with African slaves. The slaves fought back using something called small pox. Although it was primarily by accident,

Proposed Transylvania Colony

small pox began to decimate the Americans armies because all of the black slaves were bringing them their food. They were also bringing them their supplies and doing all the heavy lifting. So, the Americans they were serving suddenly got sick with small pox.

Based upon the history learned from Cotton Mather, George Washington used variolation by taking small pox from open running pustules of the troops who were infected and then making a small paper cut on those uninfected and thus infecting them in the process. Most came down with a small case and developed an immunity to it. This was magic in those days, and if it had not been for vaccination, Washington would not have been able to win against the British because all of his troops who were not dying were deserting, but he was able to save most of those who were dying with variolation.

This was all the Americans knew about biological warfare at the time and this is what they used against the American Indians. As the story runs yet deeper and what I will be offering is prone to collide with so much propaganda that you have been fed over a lifetime, it is advisable that I offer context as to the reason that you do not hear about these things. This has everything to do with the fact that the Department of Defense, who I worked with for ten years as an expert in documents destruction, runs your entire information system.

In 1994, the Department of Defense (through DAARPA, NASA and the Digital National Science Foundation for the Digital Library Initiative) went to Stanford University, not too far away from the Presidio, and took a couple of students, Larry Page and Sergey Brin, and paid them a multi-million dollar contract to develop a search engine. In 1998, it was introduced as Google, the web address for which is *google.stanford.edu*, a Federal Government address. And what does Google own? They own Google Chrome, Picasa, Google Earth, Google Mars, Google Books, youTube, the Cloud, Android, gmail, and so on. The woman who ran Google went on to run Yahoo so it is the same "Data Mafia" that is running the various search engines, and this is where you get your information. Using third party advertising, the Government is selling you this information to you as a consumer where you think you have a choice, but you do not.

To put this into an even deeper and military perspective, the men with the gold are not those who makes the rules, but rather, it is the men with the guns. I am referring to my alma mater, the Department of Defense. If you take all of their staff and the various arms of the military, they are less than one percent of the American population and they get over fifty-one percent of your discretionary annual budget without fail every single year, and this is by Congressional vote. How did it get to this point? We have to go back once again to the years before World War I and revisit Jack London, an extreme racist and ardent communist who was actually a member of the Communist Party.

London's *The Unparalleled Invasion* was based upon a pathologically xenophobic doctrine in vogue at the time of what is called anti-Siniticism. Sinitic refers to the Chinese or Vietnamese and other people who speak the Sinitic languages. This concept is well understood in Asia and the British fostered such sentiment by referring to the Chinese as the Jews of Asia as they pretty much dominated petty capitalism throughout the Pacific Rim. In order to exterminate all of this, Jack London recommended the unlimited use of biocidal weapons. His story was published worldwide in 1910 in *Collier's* magazine, one of the most well read and well distributed magazines in the entire world, it being translated into multiple languages, including Japanese. This was an open call to ethnic genocide of over 25% of the world's population. At that time, China had one billion people. That was 1/4th of the world's population and Jack London said we had to kill them all. Once that was done, we could move in and establish American democracy over the land of the dead. London's story is out there in pdf format if you want to read it.

At the same time as this magazine article by London was published, Asia was experiencing the Manchurian Plague. The great Manchurian Plague was unleashed when a mysterious outbreak occurred that was the result of a virulent virus transferring from marmots to human beings. At that time, Manchuria was the headquarters of the Ching dynasty, the dynasty that had administered China for hundreds of years. Suddenly, all of the Manchurians began to die off. When the communists took power, there was a second wave of extermination against the Manchurians. According to data from UNESCO, there are 19 native speakers of Manchu out of a total of nearly 10 million ethnic Manchus. Currently, there are several thousand who can speak Manchu as a second language by reason of governmental primary education or free classes for adults in classrooms or online. It cannot be denied, however, that one of the major factors in this language being almost lost was the huge plague that occurred in 1910.

Although no one has ever been able to prove it, many local Asians and Emperor Hirohito himself knew, as a very young man, that the Manchurian Plague was generated by artificial means. The Americans had launched it from Durham White Stevens, a career diplomat who had been set up to serve as the "American Dictator of Korea" to serve the interests of Edward H. Harriman, director of the Union Pacific railroad. Harriman wanted an Asian round-the-world railroad that would be controlled by American stockholders. For this, it would be necessary to acquire the South Manchurian Railroad, an asset Japan had acquired as a result of the Russo-Japanese War.

As 60,000 Manchurians had died within a year's time, the realization of this by Hirohito sealed the deal on his decision to become a marine biologist. He was going to defend Asia from America by using biological weapons of mass destruction.

America's proven track record of biological warfare would surpass itself during World War I, and it was the only reason the Allies would be able to claim a victory. It all began when the most powerful person in the history of American medicine, William Henry Welch, approached the 22nd Surgeon General, Major General William Gorgas, and told him that President Woodrow Wilson, who was very eager to get in on this war, could not possibly win, at least by conventional means. The reason for this is that the Germans had virtually won the war already. This is something that does not go into any popular books, but if you look it up, you will find out that what I say is true.

You can look up the Brest-Litovsk Treaty which came into being after the Bolsheviks had defeated the Czar's army, forcing him to abdicate before they eventually annihilated his entire family. As the new Bolshevik government was too weak to fight the Kaiser, they surrendered one third of their population to the Germans, twenty-five percent of their land, and half of their industries.

At that point, the Kaiser was invulnerable. He had won everything Hitler would eventually desire, and with that, he was mobilizing his troops to move into the West. If American expeditionary forces would have tried to stop this, they would have served as nothing but speed bumps in the way of the Kaiser's powerful military machine. The Americans, or Woodrow Wilson to be more specific, was in a conundrum. He badly wanted to get involved in this war but had no chance of defeating the Kaiser's powerhouse military.

It was John D. Rockefeller and his Institute for Medical Research, now Rockefeller University, who would provide the solution and means to end the war. He told Wilson that the aforementioned Welch and Gorgas had been working at his facility and were studying the history of how George Washington used variolation during the Revolutionary War and that they could produce a disease that would win the war in a manner that he could redraw the map of Europe any way he wanted to.

Taking Rockefeller's advice, Wilson signed an executive order to facilitate such. Welch and Gorgas went to San Quentin, conscripted a bunch of convicts and put them into unrefrigerated box cars to go to Fort Lewis (Washington), Fort Leavenworth (Kansas) and also various

MAJOR GENERAL WILLIAM GORGAS

other facilities in Canada, thus creating what is known as the American Army Flu. It is actually more properly referred to as the American Army and Navy Flu. The first wave was in 1918 and was the Army Flu. The second wave, far deadlier, was the Navy Flu and it paved the course for ending the war. All in all, this flu represented over 200,000,000 deaths worldwide and 20,000,000 in the United States. The absolute horror of this crime is that it changed the course of human evolution, and you can read more about this in the following books: *To End All Wars — Woodrow Wilson and the Quest for a New World Order* by

Dr. William Henry Welch

President Woodrow Wilson

Thomas J. Knock and *Lenin, Wilson, and The Birth of the New World Disorder by Arthur Herman.*

All of this was on record with the Letterman Army Medical Center where I, as a Department of Defense Research Librarian, had to collate and destroy documents. The Letterman Army Medical Center later became the property of George Lucas's Industrial Light and Magic Company after the entire Presidio military base was closed due to a number of scandals. Not only were the Letterman Army Medical Records destroyed but so was the Letterman Army Medical Center itself. The property now belongs to Disney. The reason this is so significant is not only because the Presidio served as the Pentagon of the United Nations, but the Letterman Army Medical Center once trained twenty-five percent of all U.S. Army medical doctors, and they were all trained in a military fashion. When you become a doctor in the Army, they pay for your medical education, and as a military doctor, you are immune from the Hippocratic Oath. The Surgeon General is not subject to the Hippocratic Oath either.

So that you better understand the American Army and Navy Flu and influenza in general, all flus are avian flus, all flus are swine flus, and all flus are Asian flus. The reason for this is that all flus come out of Asia, such as the Hong Kong flu, the Tokyo flu, the Taiwan flu, and so on. Why is this? It is for the reason that the majority of Asia's population is living off of subsistence agriculture, and they actually live with pigs and chickens. The reason that all flus are avian, however, is that the entire biomass of all birds in the world is a vector for influenzas that do not hurt them because all of these viruses evolved in the stomach linings of dinosaurs millions and millions of years ago. As the dinosaurs evolved into birds, the entire flight of birds throughout the world is a biomass for the evolution of influenza. All nine billion humans are sloppy seconds. We are a fly speck on the windshield of viral life, but when they do not have birds, they will take us. The virus transfers to us through pigs because the pig is so similar in chemistry to humans, and we catch diseases that they catch. That is why all flus are Asian and why they are all swine flus. They all come out of subsistence agricultural systems. This is also why Asians bow and do not shake hands. By doing so, they do not pass on the flu.

It was the Americans who created their own flu, right here in the United States. The convicts in the unrefrigerated box cars were sent to Kansas with live chickens that they were told they were going to eat. The chickens, however, "ate them" first. They all had fecal matter, droppings, and shedding feathers, creating histoplasmosis in all of the convicts, and by the time they got to Kansas, they were all infected with H1N1 (H = Hemagglutinin and N = Neuraminidase) which allows the virus to bind to and enter a specific cell.* This is usually what your immune system targets, so changes to this molecule allow the virus to avoid detection. This was a new flu, artificial, invented, and never before existing on the planet.

They found a male "Typhoid Mary", Albert Gitchell, who was immune to the disease but was a carrier, and they made him the cook. Stationed at Fort Riley, Gitchell was the first reported case of H1N1, and it was recorded on March 4th, 1918. He was cooking for thousands of men, and they all contracted it as result of being forced to sleep together. By the time they were sent to Europe, the majority of them were dying. These infected troops initially arrived in Spain, a neutral country. The Americans were able to use Spain because they had paid reparations to the Spanish government for all of the colonies they had stolen from them in the Spanish American War. These reparations allowed the Americans to land, and when they did, the ships were referred to as "coffin ships", the name deriving from the fact that most of the soldiers arriving in them arrived dead. It is by reason of these ships landing in Spain that the American Army and Navy Flu became known as the Spanish Flu.

Every medical bureau around the world, and of course their respective governments, knew that this was an American flu because every flu in the history of the world came out of Asia, and that includes the Black Plague, a pestilence which came out of Mongolia and wiped out two-thirds of the European population. All of this was due to the Mongol Empire and the spread of the giant Asian gerbil, that being the actual carrier, not the European black rat. All of this has been proven. Every scientist or biologist of the day knew this. Every government knew this. The only people who did not know this were the American public, and they died in the millions.

The American and Navy Flu was the equivalent of a nuclear war in modern times that would kill over 100 million people. For President Wilson, it was like a sacrifice to get what he wanted. Over twenty million Americans died and they all died so quickly that they ran out of coffins in the first week. There are cities with tens of thousands of unmarked graves throughout the United States. They had to bury people using pile drivers because they did not have enough men left alive to do the job. They had to actually use what could be considered as the bulldozers of those days.

As the American troops got into the trenches, the highly infectious influenza spread to their allies and enemies alike, unleashing an

---

*The handwritten notations on the documents I read referred to this as "chicken rul" or "chicken rull".

unstoppable virus that made conventional fighting by either side utterly irrelevant. The Kaiser's highly effective war machine had been stopped, but not by military might, unless you consider biological warfare to be an arm of the military, and now it was.

The success of this from a military perspective was why the Government set up the Strategic Armed Military Services or SMAS, the high command military indoctrination center where they had originally concocted these plans to win World War I. This was also how they were planning to win World War II, but we will address that in the next chapter and later learn how Hirohito beat them at their own game.

Woodrow Wilson was hailed for his leadership during World War I, a war he declared would be "The War to End All Wars", and he was given the Noble Peace Prize for his role in ending the war. His true role, however, was completely glossed over by the stupefied public. His fate, however, was not so positive. When he went to Paris to negotiate the *Treaty of Versailles*, he became struck with the very influenza that he had perpetrated against so many innocent souls. He became so ill that he was mentally deranged afterwards, becoming obsessively paranoid and concerned about petty details that did not matter, even to himself. Wilson subsequently suffered a stroke and remained paralyzed for the rest of his term in office, his wife making the major decisions for the country.

What the Americans had won would haunt them for the rest of the century. Their enemies most certainly did not forget the American Army and Navy Flu and what Wilson had wrought on the population of Europe, to say nothing of his own citizens. As a result, Adolf Hitler and Benito Mussolini made it mandatory for everyone to perform the Roman Salute instead of shaking hands, and this was in order to avoid making physical contact so as not to contract the disease that had originated in the United States and was signed into existence by President Wilson.

The result of World War I was that Woodrow Wilson's brain child, the League of Nations, was incorporated into the *Treaty of Versailles*, along with other treaties with the defeated Central Powers. Wilson, however, was never able to convince the Senate to ratify that treaty or to allow the United States to join the League. As already alluded to, Wilson suffered a severe stroke in October 1919 and was incapacitated for the remainder of his presidency.

This is actually the way World War I ended. The largest pandemic in world history was originated and sourced out of the United States, disembarked and unleashed by Thomas Woodrow Wilson, the most evil man in all human history, and NO ONE knows this, realizes it, or recognizes anything about that pandemic, not even anthropologists or epidemiologists (scientists who study the incidence, distribution and control of diseases). This is hidden history everyone needs to be aware of. You can thank your governments for hiding this from you for the past one hundred years. People are unaware there was an American war front in World War One where 20,000,000 Americans lost their lives, and anyone who denies these facts and realities are simply denying reality itself, either in ignorance, cognitive dissonance or total deniability.

Wilson Wilson brought "peace" to Europe by mass-murdering 200,000,000 people worldwide, on whose demise he redrew the map of Europe and created countries such as Poland and Czechoslovakia, all in a quest to end all wars. This, of course, did not end all wars because it actually guaranteed a Second World War, evident when both the Soviet Union and the Third Reich invaded Poland in September of 1939. The National Socialist Party came into existence because of the way World War I had ended and ticked off a group of political zealots called the National Socialists or German Workers Party. If the Americans had not ended World War I by unleashing the American Army and Navy Flu, there would have been no Nationalsozialistische Deutsche Arbeiterpartei (NSDAP), and therefore no "Endlösung der Judenfrag" or "Final Solution" to the Jewish question. Genocide begets more genocide.

While Wilson never recovered and lived only a few years after the end of the war, he had a protege who would carry on with his plan for a New World Order.

CHAPTER FOUR

FDR

A favorite of President Wilson, Franklin Delano Roosevelt campaigned hard to help him win the 1912 presidential election. Wilson rewarded Roosevelt by appointing him as Assistant Secretary of the Navy, a position he relished and from which he waged war as aggressively as did the President himself, if not more so.

Although Josephus Daniels officially held the title of Secretary of the Navy, he was largely bed-ridden and struggling with senility. During World War I, his duties were relegated to handling policy and formalities while his top aide, Franklin Delano Roosevelt, handled the major wartime decisions.

Like his distant cousin, Teddy Roosevelt, Franklin Roosevelt had ambitions to become President, exhibiting a strong appetite for war and was hungry to prove himself on the battlefield. Keep in mind, he was not yet a disabled individual, and in those days, you were supposed to have military experience in order to serve as President.

FDR was ambulatory, mobile and was part of the process for America's plan for worldwide conquest as the country developed what was called the Rainbow War Plan, it being named such as the Americans designated every single country they wanted to go to war with by a different color.

The plan for war against the British Empire was War Plan Red, and this was motivated by the latter's hold on the incredibly lucrative opium empire that had been the primary source of British wealth for hundreds of years. While the color red represented the British Empire, War Plan Crimson was a subsidiary plan of War Plan Red and was a war against Canada. Crimson was the most specific and the most detailed of all the war plans because FDR had every intention of realizing it and had expanded it voluminously by putting tremendous energy into it. We will address this further as it is crucial to understanding FDR's motivation for World War II, but first, I will give an account of his exploits during World War I.

Although it is not well recognized, the battle front of World War I extended to North America, and more specifically, the country of Haiti. Germany had a small presence on the island, but one that was considered both significant and menacing. Haiti was the only country in the Western Hemisphere with a black government and foreigners were not allowed to own land. Certain Germans had gotten around this law by marrying Haitian women, and America justified this as a potential threat. The U.S. therefore undermined Haiti's economy by seizing Haitian assets in American banks and then sent in an invasion force led by the Assistant Secretary of the Navy himself, Franklin Delano Roosevelt. It was an occupation that lasted nearly twenty years.

Roosevelt was directly responsible for some of the most bloody massacres of blacks in the history of humanity. He killed blacks and tortured them in the most horrendous manner, employing gratuitous sadism to get information. This included Haitian's being roasted on spits and it earned Roosevelt the name of the "Butcher of Haiti".

After the war, Roosevelt ran for Vice-President on the same ticket with James M. Cox, but his past caught up with him. Although their opponent, Warren Harding, was a known member of the KKK, he was able to defeat Cox by pointing out that FDR's track record of killing blacks and torturing them was a sure indication that they would be better off voting for Harding than Cox and Roosevelt. Harding was able to get the black vote and win because Roosevelt was that bad in terms of his scandals against blacks in Haiti. A direct quote from Warren Harding is as follows.

> "Practically all we know is that thousands of Native Haitians have been killed by American marines, and that many of our own gallant men have Sacrificed their lives at the behest of an Executive Department in order to establish laws drafted by the Assistant Secretary of the Navy ... I will not empower an Assistant Secretary of The Navy to draft a constitution for helpless neighbors in the West Indies and jam it down their throats at the point of bayonets borne By U.S. Marines."

Like Wilson, Roosevelt's karma soon caught up with him when he contracted polio and became paralyzed for the rest of his life. This did not, however, deter his political ambitions, and when he ascended to the Presidency, he set out to revive the Rainbow War Plan and set out to conquer the world. Japan was very much on the horizon, and the plan for that country was War Plan Orange, and so it was for each targeted country, each one represented by a different color.

Although it does not reach the popular press, world conquest is a continual game played by the players for power in society. It is not new to the world, and it is something every U.S. president or world leader has to deal with. This mentality, which Emperor Hirohito realized very early on in his youth, is one of conquer or be conquered.

**BUST OF FRANKLIN DELANO ROOSEVELT
"THE MAN WHO WOULD BE KING"**

The British opium trade into China was the most lucrative trading operation in the world, and it was hundreds of years old. The United States had been trying to gets its share of Chinese trade ever since Caleb Cushing was sent to Asia as the first American envoy to China in 1843. The British were squeezing the Chinese, forcing each Chinese citizen to purchase a periodic allotment of heroin. If you can consider a billion people purchasing a monthly stipend of opium, you can imagine the tremendous amount of money this would generate. China was Britain's cash cow and literally kept the Empire afloat. In Europe, however, the position of the English, particularly the Crown, was not so secure in the decade prior to World War II.

Years before the invasion of Poland in 1939, Germany was already viewed as a menace to the interests of England, and the Crown was especially vulnerable and sensitive to this fact. Just as many Germans fled to Argentina at the end of the war, the Royal Family had plans to escape to Canada in the event of an invasion.

With regard to the war plans he had devised as Assistant Secretary of the Navy, Franklin Roosevelt set his sites on Canada as his first step. This would bring the United States directly into war with the British

FDR'S WAR PLANS WERE LEAKED IN THE
CHICAGO DAILY TRIBUNE OF DECEMBER 4, 1941,
THREE DAYS BEFORE THE PEARL HARBOR ATTACK

Empire; but more importantly, it would cut off the escape route of the Royal Family. For Roosevelt, the end prize was the sequestration of the opium trade, an enterprise that was directly controlled by the Crown.

As he conceived it, Roosevelt's War Plan Crimson called for the unrestricted deployment of poison gases and the strategic bombing of Halifax. Roosevelt submitted actual plans to his American commander to invade Canada and seize Halifax and Montreal in a lightning campaign before the British could show up and intervene decisively. The deployment of poison gas was authorized to be used from the inception of imbedding hostilities. It would be a war against humanity itself where populations themselves would intentionally become the front lines. The entire purpose was to leave no Canadian alive. This was a war of genocide he was actually planning. The Americans would repopulate a depopulated Canada, and the entire country would all be Americans.[*]

In February of 1935, the U.S. Department of War arranged for Congressional appropriations of $57 million dollars to construct three air bases on the border with all haste for the express purpose of conducting preemptive surprise attacks on Canadian air fields. The base in the Great Lakes region was deceitfully camouflaged as a civilian airport but

---

[*]You can find this out in the U.S. Military history collection stored at Carlisle Barracks in Pennsylvania. The code I remember before burning all of the documents was AWC2 1936-1938, G2 code #19A. You can still verify certain aspects of War Plan Crimson in regular history books and even on the internet.

was marketed as being capable of dominating the industrial heartland of Canada: the Ontario peninsula.*

Based upon their hundreds of years of Imperial experience, the British well understood the advantage of naval control and the fact that North American geography provided its own inland ocean. They had their own plans. The Great Lakes would serve as the shield of Canadian dominion, another sword by which to cut and bleed the United States during the period of Anglo-American tensions that would ensue, both sides making actual military preparations for a war that never happened. This included the building of naval forces for the Great Lakes.

If war were to break out, the powerful British navy would threaten American interests and control in the Philippines, Alaska, Hawaii, Guam, Samoa, Cuba and Puerto Rico as well as the Panama Canal. In exchange for all these losses Roosevelt was willing to absorb, all of which were calculated and even expected, he planned to conquer Canada and cut off the escape of the Royal Family in order to secure the lucrative opium trade.

In August of 1935, over 36,000 regular United States Army and United States National Guard Troops, elements of five divisions, maneuvered as the First Army in a show of force that anticipated commencement of War Plan Crimson, a first step in the context of socialist America's Rainbow War Plan. Canada would be invaded, even if they declared neutrality, with the Canadian government being abolished and the conquest held in perpetuity. It would be made clear that she would suffer grievously for any resistance.

This waving of the flag across all of upstate New York was backed up by yet 15,000 more U.S. National Guardsman conducting core level operations in Pennsylvania. It was a tangible threat of invasion that would deprive the British Royal Family of the dominion of Canada by way of optimal evacuation in the event of an invasion of the British Isles by the Third Reich; and it was ultimately the threat of imminent conflict with National Socialist Germany that forced the Royal Crown to stave off American aggression against their commonwealth by forfeiting the lucrative opium trade with China to Roosevelt's socialist administration.

That was the original war on the Red Empire (Great Britain) that was planned by Roosevelt down to the minutest detail. Although it never resulted in all out war between the two counties, the ominous threat of a German invasion in Europe coupled by the menacing military presence of the United States on the Canadian border literally forced

---

*These references are electronically accessible on microfiche through the United States National Archives. People can also look up FDR's 1935 plan for the invasion of Canada and can do their own research on this.

TYPICAL HILLY TERRAIN IN MAINLAND CHINA

the hand of the British, thus enabling the United States to acquire the opium trade in Asia. This prize would become an arsenal to totalitarian democracy's manifest destiny and an industrial platform from which to rebuild and eventually regain all that would be lost elsewhere. For the United States, it can be considered that it was and still is a virtually bottomless slush fund for black operations.

Besides the menacing overtures from the United States, there was another factor which prompted the British to acquiesce to Roosevelt's demands. Completely independent from the United States, Britain's opium trade was also being challenged at that time. In the first place, it was very difficult to import anything into China as the hilly terrain made highways non-existent. It all had to be done via rivers. At that time, most Chinese lived their entire life within a parameter of ten square miles. Not only were such villages and population centers hard to access, these Chinese territories were controlled by war lords. These factors, however, were not the only problem the British were facing with their opium imports.

The Japanese had been very economically challenged due to outright restrictions or tight controls being placed upon them by European colonial powers. To overcome this, they were making peace with the Chinese to stimulate trade for their own purposes. A doped-up China was not good for the Chinese economy and would make them a difficult trading partner. Accordingly, the Japanese made things difficult by interfering with the abundant proliferation of opium into China, an enterprise they were not interested in for themselves.

When England accepted the proposition of maintaining the dominion of Canada in exchange for the lucrative drug trade, it gave Roosevelt a problem he was ready, willing and able to overcome. Instead of relying on rivers for transportation, Roosevelt's response was to create his own private air force, a unit which could fly in the opium and bypass all of the difficulties on the rivers. This air force eventually became known as the Flying Tigers, and later, Air America, an outfit that has a notorious reputation in and of itself.

Before we can truly understand and appreciate the heritage and the circumstances by which Franklin Delano Roosevelt was able to muster his private air force, completely out of the boundaries of ordinary legalities, we have to understand more about his background and political leanings. While no one will dispute that he was an ardent socialist who took the United States off the gold standard, he was also passionate about communism, as were the marines he chose to secure and maintain the in-theater airfields that rendered the aforesaid extralegal air force and its criminal enterprises operational.

THE ROSWELL DECEPTION

CHAPTER FIVE

COMMUNIST PEDIGREE

Although it is neither broadly known or historically appreciated, the United States invaded Soviet territory in 1918. This invasion amounted to an actual series of battles between the Russians and Americans which was – and still is – obliquely referenced as the Siberian "Intervention", "Expedition" or simply "Incident". It is always obfuscated by such obtuse terminology because no one dares admit that the Americans invaded the Soviet Union*, but the Bolsheviks had taken over Russia during that year and all legitimate governments recognized violent Communist insurgency as a major threat. Woodrow Wilson responded immediately by sending in the U.S. Army in what became the only recorded succession of massive engagements in the field between the U.S. military and the Soviet Union, all of it within Soviet territory. Internationally coordinated as part of an even grander-scaled "Allied North Russia Intervention", the Siberian invasion by the U.S. was accompanied by their great ally, Japan, who fought side by side with Americans against the Soviets.** To put these events in context, it is important to review the timing of both Vladimir Lenin and Woodrow Wilson on the world stage.

Lenin came to power in October of 1917, in the same year Wilson began his second and final term in office. Their respective rises precipitated what would become known as the Cold War, but circa 1918-1922, the confrontation was both openly direct and hotly combative between their two very different ideas of collectivism: Lenin's communism and Wilson's racist internationalism.

The Russian Revolution and a mutiny in the French Army convinced the United States that Russia and France would pull out of the war, leaving the way open for a German victory. This was unacceptable to the Wilson administration. These two events drove Wilson's decision to ask Congress to declare war on April of 1917, only the second declaration of war in U.S. history.

---

*Technically speaking, the government that became the Soviet Union was then recognized as the Russian Soviet Federative Socialist Republic, a designation that lasted from 1917-1922.
**This is with the exception of several incidents where cultural miscommunications resulted in direct Japanese-American combat, such as what historically transpired in Evgenevka and most outstandingly on the First of October in 1919 when the Japanese sided with the anti-communist Russian White Army in defense of a fellow monarchy.

Lenin and Wilson died within days of each other in 1924, a cosmic coincidence with intriguing parallels. Becoming convalescent recluses toward the ends of their lives, both had repudiated capitalism, even to the extent that Wilson began expressing that the rising fear of "coloreds", immigrants, radicals, and anyone perceived to be a Communist threatened to rupture the fabric of American unity.

It was only after Wilson had become an inert president, suffering a stroke to the point of disability, that those troops from Siberia were withdrawn, all of which led to the ultimate resurgence of the Soviet Union. It was after this that Franklin Roosevelt decided the Soviets were the wave of the future, and he decided that he was going to side with them.

Roosevelt had followed communism from the very beginning and was rather obsessed with it, and it is likely he became a communist adherent during the time of Lenin and well before Stalin took power. Ordinary history will tell you that Roosevelt was an ardent Stalinist and did everything he could to support him prior to the war and during it. When Roosevelt died, Stalin was dismayed and said, "President Roosevelt has died but his cause must live on. We shall support President Truman with all our forces and all our will."

What is probably the most damning evidence of Roosevelt's alliance with communists is his choice of Lauchlin Currie as his White House Economic Adviser. Currie was a Canadian citizen and a Marxist economist from Harvard who was associated with other Marxists. Currie not only drafted new laws for the Federal Reserve in order to strengthen its power, he completely captivated and manipulated Roosevelt on anything to do with economics.

It was the FBI who exposed Currie as a spy during this time period. This was the result of the VENONA program, an ongoing project that decrypted Soviet cables before the war and lasted throughout it. These cables identified Currie beyond all doubt as a source of Soviet intelligence and exposed him as operating under the cover name of "PAGE".

LAUCHLIN CURRY

Soviet intelligence archives identified him under the code name "VIM". These were exposed after the collapse of the Soviet Union when public sources were given access to all Soviet records until Vladimir Putin came into power and reclassified all former Soviet documents.*

The FBI's investigation included records of personal dispatches from Currie to Stalin informing him of how he was destroying America's economy and manipulating Roosevelt. When this information was brought to Roosevelt himself, he acted to protect Currie as opposed to hanging him for treason.

Roosevelt's communist leanings did not go unnoticed. Most concerned were powerful business executives from corporations like J.P. Morgan, the DuPont dynasty, Goodyear, the American Legion, owners of the other multinational companies and even some former presidential candidates. These are the people who formed the inner circle of what became an underground movement to overthrow Roosevelt's government. These business leaders were not only very displeased that FDR had abolished the gold standard but that he was hellbent on nationalizing their own industries in the Stalinist mold; and further, that he was using these precepts for mobilization of a huge war.

Roosevelt's polio became so severe that it was painfully obvious to anyone who dealt with him on a face-to-face basis, and this included the aforesaid business leaders, that it was impacting his mentation to the degree that he could not adequately handle the running of the Government, let alone deal with the combined problems of the Great Depression and foreign enemies. He was abusing the Office of the President by using increasingly dictatorial powers in an attempt to overcome the Great Depression.

Just as Roman senators had chosen Brutus to be the point man to spearhead the assassination of Julius Caesar, so did these industrialists seek out a character who could serve the purpose of leading their anti-communist crusade against FDR. Accordingly, they approached Smedley D. Butler, a retired Marine Corps Major-General and decorated hero of World War I. To them, Butler appeared a sound choice in national loyalism: he was a war hero, had won the Medal of Honor twice (a

---

* The VENONA intercepts and the KGB notes of the NKVD, the predecessor of the KGB, characterized Lauchlin Currie as the head of the Silvermaster Spy Ring of Soviet agents and it was called such because they manipulated FDR into taking America off the gold standard. Currie's cover was later publicly blown by the admitted Soviet agent, Elizabeth Bentley, in August of 1948. Currie appealed to the United Nations for international asylum, and he was consequently appointed in 1949 to head the first of the World Bank's comprehensive surveys in Columbia where he ultimately stayed on as a Columbian citizen in order to integrate Columbia's economy into the world banking system.

SMEDLEY BUTLER

statistical impossibility as there are only eighteen other living recipients in the annals of American Military History who won the medal once), and commanded intense loyalty from many soldiers and veterans.

Their strategy was for Marine General Smedley Butler to deliver an ultimatum to Roosevelt demanding that he step down: first, that he surrender the day-to-day governance of the United States to himself (with the backing of American Capital) or face the prospect of a military coup, supported by veterans of the American Legion.

Roosevelt himself would be encouraged to honestly acknowledge incapacitation from his polio to his taxpaying electorate and responsibly relinquish power. The conscientious industrialists hoped that President Roosevelt would create a new cabinet position, the Department of the Secretary of General Affairs, which would be headed by Smedley himself, who would run the Government as a sort of regent according to the economically sound advisements of the industrialists. This Department of General Affairs would have reversed many New Deal policies, in accordance with the wishes of Wall Street. For example, it would reinstate the gold standard for currency, a regulation which Roosevelt had abandoned in order to subvert the American economy toward collectivization. In the end, General Smedley Butler would become the functional (èntendrè) ruler of the United States with Roosevelt reduced to a figurehead position.

The free marketeers understood that the only way to prevent America from becoming a communist satrap while, at the same time, being able to maintain a competitive par with the depression-proof Fascist economic engines along the lines of Germany or Italy, was to institute a monopolist government with Smedley in charge and Big Business giving him his orders.

In the event that FDR would not play ball with their scheme, they planned for General Butler to lead an army of half-a-million (500,000) veterans and force the President from office. Robert Clark, one of Wall Street's richest brokers, pledged $30 million (over $588 billion by today's valuation in 2021) toward regime change, secretly swearing by his life that "America Must Make Business, Not War – For War Is Not Our Business." Such individual financial commitments, as opposed to corporate or conglomerate sponsorship emphasizes the immediate existential threat posed by Roosevelt, a bona fide Russian collaborator.

To put this in the proper perspective, it is very relevant to grasp the fact that this was during the time of the Great Depression, and these industrialists were virtually the only few Americans who were contemporaneously employed or were employed at all. Together, they essentially constituted the American economy itself at that point in history.

At the complete opposite end of the moral spectrum from the industrialists, General Butler himself hard-bargained the situation so as to astronomically expand his own personal fortune. In other words, he accepted a substantial bribe, and verbally acquiesced to their demands, or so it seemed. This, however, was a huge mistake and miscalculation by the industrialists. They perceived General Butler as the greatest epitome of American honor because he was one of a rare group of individuals who have been awarded the Congressional Medal of Honor, America's highest award for military valor in combat, but he had actually won it twice. So that you understand the rarity of such a circumstance, understand that you practically have to die as most of these awards are given posthumously. Accordingly, they thought he was the man for the job.

Had the industrialists done their homework, they would have learned that it was actually Franklin Roosevelt himself who had proposed Butler for his second Congressional Medal of Honor, as a result of his combat success in subjugating the population of Haiti. Of further relevance was that Smedley Butler was an avowed dyed-in-the-wool communist who spoke frequently at communist rallies and was enamored with the vision of a worldwide communist revolution. This included the doctrine of fomenting a global war that would sunder all international order and usher in a universal revolt by the proletariat working class. Butler was completely dedicated to the destruction of the United States itself as he wanted complete communization.

After taking their money, and instead of taking on the role of a true American hero, Butler ran straight to the media and all his friends in Congress and squealed, avowing his dedication to Roosevelt. Had Butler accepted the backing of the industrialists, they were indeed powerful enough to have put him in power and quell the media at the same time. Butler's squealing, however, resulted in the McCormack-Dickstein Committee Hearings.* They officially listened to Butler's testimony, and the press, who referred to it as the "White House Putsch", first reported the alleged plot earnestly but soon changed course when a *New York Times* editorial suddenly dismissed it as a "giant hoax". The hearings went nowhere as all the hot issues were covered up. The reason for this was not only because of the industrialists' influence but by the fact that the Congress knew that if they dug deep enough, it would betray the President's own treason and other communist influences which necessitated the conspiracy in the first place. The official report edited out the names of the most powerful "Gold Standard Confederates" (referring to the industrialists), and to this day, the transcripts of the hearings cannot be traced publicly. The industrialists avoided both scrutiny and prosecution.

In the wake of these impotent hearings, Butler took to the airwaves to promote himself as "America's Man On Horseback (or 'National Savior')" but found no interested audience as he offered no one any jobs and communicated nary a single solution to the lack thereof. In the end, the Free Market Front failed in their united effort to preempt America's provocation of the next World War because, despite everything, Smedley was himself ideologically devoted to communization of the United States.

To many Americans, both Roosevelt and Butler are American heroes, but too few realize their actual pasts and prejudices. So, while Roosevelt was deeply in bed with communists such as Smedley Butler and Lauchlin Currie, he had another key communist marine up his sleeve that would play a key role in precipitating World War II. That

---

*With regard to the McCormak-Dickstein Committee, it was subsequently reorganized into the House Un-American Activities Committee. While John McCormack went on to become Speaker of the House from 1962-1971, Samuel Dickstein earned himself a reputation as a Soviet agent whose handlers referred to him as "Crook". He was found to be on the payroll of the NKVD (the predecessor the KGB) for $1,250 per month.

was Evans Fordyce Carlson. Highly decorated by the Soviets upon his death for his service as an honored agent of the Soviet Union, Carlson is a legendary member of the U.S. Marine Corps. The legendary leader of "Carlson's Raiders" during World War II, Evans Fordyce Carlson worked arm-in-arm with Mao's communists and developed the model for the modern Marine Corps, its status and it special forces, even popularizing the phrase of "gung-ho". Besides being a card carrying communist, Carlson was also an outspoken member of the Communist Party and played a key role in the events that would precipitate an all out war between Japan and the United States. Before we address the antics of American communists in bringing about World War II, I will share a short history of the shaping of the Marine Corps.

Evans Fordyce Carlson

CHAPTER SIX

THE MARINE CORPS

The original purpose of the USMC (United States Marine Corps) was their ability to carry out amphibious landings on hostile beaches. Even more proficient than their capability for such is their capacity for public relations. The truth is, the US Army conducted the biggest amphibious assault in our nation's history when they captured the Normandy beaches, and neither the Army nor the Marines have assaulted an enemy held beach since the Korean war, some seventy years ago (at time of this writing). In every subsequent conflict, soldiers and marines have fought in the same way, using similar equipment and tactics.

The Marines are, in fact, a second and rival Army, and since they compete with the Army for funds, missions, and prestige, their real enemy is and always has been: the US Army. The Marine Corps, however, has an unfair advantage in this competition: the National Security Act of 1947 which prevents any changes in the force structure of the Marines. The fact is, for most of their history, the United States Marine Corps was little more than a security force for the Navy.

The myth of the Marine Corps as a second (and supposedly superior) army began in WW I. When the United States entered the war in 1917, over two million U.S. Army soldiers were deployed to France along with one brigade of marines, about ten thousand strong. Despite being a tiny fraction of the American forces fighting in WW I, the Marines managed to make a name for themselves at the U.S. Army's expense.

General Pershing, the Commander of all U.S. Forces in France, had ordered a news blackout that prevented reporters from mentioning specific units in their dispatches. The purpose of the order was obvious; to prevent German intelligence from learning about American troop movements. One reporter, however, circumvented the order, and he was a war correspondent for the *Chicago Tribune* named Floyd Gibbons.

After Mr. Gibbons was severely wounded at the battle of Belleau Wood, the press corps passed on his dispatches without the approval of army censors. The result was a storm of press coverage in the U.S. claiming that the Huns were being defeated with "the help of God and a few marines". No mention was made of the thousands of army soldiers who were fighting and dying with equal valor. Floyd Gibbons made no secret of his "friendship and admiration for the U.S. Marines" and his prerogative served the Marines so advantageously that they posthumously made him an honorary Marine in 1941. Floyd Gibbons helped enhance the image of the Marines, but the United States Marine Corps as we know it today came of age in WW II during the Truman administration.

President Harry Truman, who played a very influential role in the events surrounding the Roswell Incident, was well aware of the adulation flaunted upon the Marines in contrast to the regular army. He viewed this as both biased and dangerous to the morale of the Army regulars. Being the only U.S. president to see combat during the First World War, Truman's perspective was the result of hard-won experience during World War I; and further, his lessons learned as a combat veteran affected the course of his life and influenced his rise to the presidency in two important ways. First, he discovered a leadership ability he had not known that he possessed, and second, he garnered a significant political base that supported him in his rise though the political ranks. Indeed, Truman was quoted as having said, "My whole political career is based on my war service and war associates." By reason of his own personal experience as a commissioned officer of the United States Army in command of soldiers, Truman was made painfully aware of the fact that their morale was sapped by a blatant journalistic bias that all but ignored their sacrifices.

Another example of this sort of bias was evident in one of the most iconic moments of World War II: the raising of the American flag at Iwo Jima. Most Americans believe that the Marine Corps won the war in the Pacific while the U.S. Army fought in Europe. In the Pacific Theater, the Navy adamantly refused to place their fleet and their Marines under the command of the Army. After five weeks of bureaucratic wrangling, General MacArthur was given command of the Southwest Pacific Theatre while Admiral Nimitz had jurisdiction over the remainder of the Pacific ocean. The result, in MacArthur's own words, was a "divided effort, the ... duplication of force (and) undue extension of the war with added casualties and cost."

The U.S. Army fought the main force of the Japanese Imperial Army in New Guinea and the Philippines. The Navy and Marines carried out an "island hopping" strategy that involved amphibious assaults on islands such as Guadalcanal and Saipan. General MacArthur complained bitterly to the President that, "these frontal attacks by the Navy, as at Tarawa, are tragic and unnecessary massacres of American lives." By way of comparison, General MacArthur's army killed, captured, or stranded over a quarter of a million Japanese troops during the New Guinea campaign, at a cost of only 33,000 U.S. casualties. The Navy and Marines suffered over 28,000 casualties to kill roughly 20,000 Japanese on Iwo Jima alone.

Even then, the Army played a greater role than the Marines would ever allow the tax paying electorate to know. Not only did the Army have more divisions assaulting Okinawa than the Marines themselves, they were consistently salvaging whatever perilous situations the Marines got themselves sunk into. This pattern was prevalent all over the Pacific Theater of Operations throughout the Greater East Asian War.

The famous image of Marines raising the U.S. flag on Mount Suribachi on Iwo Jima is actually a staged photograph of the second noncombat flag-raising ceremony. The Marines raised the flag a second time to replace the original smaller flag in order to provide the press corps with a better photo opportunity. That phony photograph has become one of the most enduring images of WW II and served as the model for the Marine Corps Memorial statue. The extent to which the Marines would go in using public relations to bolster their image has been further evidenced in recent years when they have admitted that two of the marines they identified and publicly credited in the iconic flag-raising photograph were not even involved. The significance of that photograph was evidenced by the statement of the Secretary of the Navy, James Forrestal, who was on Iwo Jima that morning in 1945. Personally witnessing the orchestrated second flag-raising of the stars and stripes, Forrestal declared, "The raising of that flag on Suribachi means a Marine Corps for the next five hundred years!"

Despite the urgings of the Secretary of the Navy, the Marine Corps was nearly legislated out of existence two years later. After the bureaucratic infighting that characterized their inter-service relations during World War II, there was a strong desire amongst military professionals to unify the military commands. President Truman agreed, and in 1946, his administration proposed a bill to unify the separate service bureaucracies. Having one budgetary authority for the armed forces and one chain of command each for land forces, ships, and aircraft makes sense. This, however, would have placed the U.S. Navy at a distinct disadvantage. As the Navy had their own air wings aboard their carriers and their own army, the Marine Corps, dividing the military would leave them at a distinct loss.

The Navy and Marine Corps were therefore determined to scuttle this legislation. Marine generals created a secret office code named the Chowder Society to lobby behind the scenes, all of this in direct opposition to their President and Commander in Chief, to thwart the unification bill before Congress. The Commandant of the Marine Corps even made an impassioned speech before Congress to plead for his separate service. It worked. Congress rejected the Truman administration's unification bill and instead passed the National Security Act of

1947. This Act guaranteed separate services with their own independent budgets, and it was a victory for the Navy and Marine Corps.

In addition, the Marines succeeded in having their separate force structure written into the language of the legislation. It is very unusual for Congress to dictate the actual composition of a military service. The National Security Act, however, mandates that the Marine Corps must maintain "not less than three combat divisions and three aircraft wings and such land combat, aviation, and other services as necessary to support them."

President Truman was furious, and military professionals were appalled. General Eisenhower characterized the Marines as "being so unsure of their value to their country that they insisted on writing into the law a complete set of rules and specifications for their future operations and duties. Such freezing of detail ... is silly, even vicious."

The war between the Army and Marines would get more vicious in Korea. On the 27th of November in 1950, a division of Marines 25,000 strong was ordered to proceed along the west side of the Chosin reservoir while a much smaller task force of 2,500 Army troops went up the eastern side. Waiting for them were 120,000 troops of the Chinese Communist 9th Army Group. The Army soldiers fought a running battle for three days against a Chinese force eight times their size, in temperatures as low as minus 35 degrees. Despite the death of two commanding officers, the task force lumbered south with over 600 dead and wounded soldiers loaded into trucks while fighting through repeated ambushes until, fatally, they were treacherously bombed by U.S. Marine Corps aircraft. Finally, just four miles from safety, the convoy was cut off by the Chinese and annihilated. Only 385 men made it to the safety of American lines by crossing the frozen Chosin Reservoir. The First Marine Division, with the help of Allied air power provided by foreign nations participating under United Nations authorization, managed to fight their way out of the Red Chinese encirclement.

Marines claimed that the Army had disgraced itself and passed on stories of U.S. soldiers throwing down their weapons and feigning injuries. A Navy chaplain attached to the Marines even made statements to the press and wrote an article accusing army soldiers of cowardice. There were so few officers and men left from the Army Task Force that the Marines' claims were accepted as fact, but Communist Chinese documents released circa 2005 prove otherwise. The Army task force fought bravely against overwhelming odds before being destroyed, and their stubborn defense bought time for the Marines to escape the encirclement. Nevertheless, the Marines to this day hold up the fight at the Chosin reservoir as "proof" of their superiority over the Army.

THE MARINE CORPS

In the context of all this, it should be pointed out that the U.S. Marines were so communized by the end of World War II that Truman feared they would overthrow him and considered them to be the greatest enemy of the United States. Upon taking office after Roosevelt's death, Truman disbanded the Marine Corps' elite units as they were the most ideologically communized. He even said that "they have a propaganda machine that is almost equal to Stalin's". That is an actual quote that you can look up.

So it is that "Give 'em Hell, Harry" Truman's cumulative combat command experiences as both a frontline officer and as Executive Commander In Chief through the torturously circuitous conclusion of the Second World War and through the Korean conflict culminated in his genuinely open contempt for the United States Marine Corps, but Truman caught hell himself for his anti–Marine Corps bias.

Truman had already undergone one of the most stressful presidencies in history, but everything spiraled into free fall after he won the election to the presidency in his own right in 1948, defeating the heavily favored Republican Thomas Dewey. The following year, a State Department white paper published in August of 1949 would reveal that China, the world's most populous country, had fallen to the communists in the very same month that concluded with the terrifying news that the Soviet Union had developed the atomic bomb a full decade earlier than expected. The year after that started with the conviction of Alger Hiss for passing secrets to the communists and Senator Joseph McCarthy brandishing a list of 205 "known communists" in the State Department during a speech in West Virginia.

Meanwhile, the U.S. Marines had never taken kindly to Truman's justifiable hostility and his understandable efforts to undermine them. At a meeting of their "Softball and Chowder Club", they conspired and decided to murder their own duly elected President. Just two weeks before the 1950 congressional elections, a hit squad of purported "Puerto Rican Nationalists" stormed Blair House, where Truman was residing while the White House was being renovated. During this period, no one but the military was aware of the First Family's relocation during the repairs.

The assassination attempt failed, but a White House policeman stationed outside was killed and another grievously wounded in a fusillade that engulfed the Blair House lobby. Had the assassins waited but another twenty minutes, they would have killed Truman and likely would have massacred his entire family as the only thing saving the life of the thirty-third President of the United States and his family was the fact that their car broke down on the way home. Had Truman not fortuitously suffered a tire blowout and had he not gotten out and fixed that tire himself only to get back into the car and drive home to find a dead law enforcement officer and a critically injured surviving constable to report the abortive mass-assassination, there would be a communist Marine Corps junta overtly running America today – in disambiguation from the covert military industrial complex doing so by default.

Enkindled by the brazenness of the USMC, that emergent military junta did not relent. Having failed to take the man's life along with those of his family, they continued their campaign of assassinating his character to a degree that would eviscerate his presidency and render himself effectively unable to govern the nation which had legally elected him. The 1950 mid-term elections, just two weeks after the assassination attempt, would see Republicans score 52 percent of the vote with Truman's own party garnering a mere 42 percent while seeing its majority in the Senate cut from twelve to two and from seventeen to twelve in the House.

By 1951, newspapers were blaring awful stories of Americans dying in Korea, spies in the Government, and corruption at the door of the Oval Office. Seven members of Truman's administration, including some of his closest aides, would eventually go to jail. Truman no longer had any political legs to stand on and chose not to run for another term as President.

This is a general background as to how communism influenced the key players who would directly precipitate a war with Japan. We will address those events in the next chapter.

CHAPTER SEVEN

DRUGS AND WAR MONGERING

After Lauchlin Currie was exposed as a communist spy by the FBI, Roosevelt protected him by sending him on a personal mission to China to aid and organize Mao's communists.

Before Lauchlin Currie came to China, the Chinese public identified the Japanese as heroes because they were helping the Americans and the British control piracy in the Chinese river system. This was absolutely necessary because there were no modern roads in preindustrial agrarian warlord China. My father was serving on the patrol gunboats in warlord China, and there were no highways or methods of transportation of any reliability other than the river system. The Japanese had their own interests in protecting this river system, but Lauchlin Currie changed all of that. The Japanese had now come directly into the crosshairs of the American drug trade.

It was during this mission to China where Currie coordinated diplomacy with direct orders to his own men in the field: Evans Fordyce Carlson, the card-carrying communist Marine Corps General discussed in the previous chapter; and Claire Lee Chennault, the founder of the Flying Tigers.

CLAIRE LEE CHENNAULT

At the time he was selected to muster up an Asian air force for Roosevelt, Claire Lee Chennault was a former fighter pilot who was "medically retired" from the U.S. Army by reason of insanity. Subsequently finding employment with the Chinese government and working for them as a civilian mercenary, he was a close ally of Chiang Kai-shek and particularly Madame Chiang Kai-shek. In August of 1937, Chennault established the American coordination of the Asian narcotics trade through the China National Aviation Corporation (CNAC), originally an opium airlift operation, via Douglas DC-3s and DC-4s.

The Japanese knew all of this opium was coming through the Americans and they were interdicting and shooting down all of these American drug shipments. As too many aircraft were being shot down, Chennault then brought in fighter craft to protect the DC planes that were airlifting all the opium, and this brought them into direct conflict with the Japanese who, at that time, were engaged in peace negotiations with the Chinese government. As was stated previously, the Japanese were establishing trade with the Chinese due to economic strangulation from the colonial powers in the Far East.

It is at this point that Currie orchestrates the services of Evans Fordyce Carlson to undermine the peace talks going on between the Japanese and the Chinese. Carlson had his American marines dress as Chinese guerrillas in black coolie uniforms in order to instigate a hostile military attack directed at the peace negotiations. The Japanese, who were the target of the attack, understandably thought that the attack was coming from the Chinese themselves and not the Americans. After all, these agitators appeared as Chinese, not Americans. The Japanese response was vehement and began with what is known as the rape of Nanjing, then the capital of China, and it resulted in hundreds of thousands of deaths.

It was only after interrogations of their Chinese prisoners that the Japanese came to realize that it was not the Chinese who were attacking them during the peace negotiations but rather Caucasians. As the Chinese being interrogated by the Japanese did not know whether the attackers were the British or the Americans, the Japanese retaliated against both by bombing the American ship *Panay* and Her Majesty's ship the *Lady Bird* in December of 1937.

For the Anglo-Americans Allies, this was the actual beginning of World War II in the Pacific Theater. Although it was recognized as a full blown war and even referred to as such by the press, the Americans in particular could not officially declare war. If they did, it would be politically embarrassing beyond belief. With respect to World War II itself, this was the first in a domino chain of potential

embarrassments that extends all the way through the Roswell Incident. The fact that it was about a black budget involving FDR's private opium trade makes it all the worse. While we often hear of atrocities committed by the Japanese, as if they all happened spontaneously, we never hear what generated the hostilities in the first place. The Americans have historically tried to hide what they want no one to know about. If the Americans had declared war at the time, everyone would soon have learned not only what they did but what they had been involved with to that point.

Over time, America would foster circumstances by which they could formally declare war against the Japanese empire and do it in such a manner that it would incite vast public support. Prior to Pearl Harbor, the United States had not declared war against Germany, a proposition that had inadequate public support due to the isolationist ideology of most Americans. This, however, became a moot point when the outright hostilities from Japan manifested at Pearl Harbor.

Despite the open hostilities between the Japanese and the Americans in 1937 and 1938, there was still the 1911 *Anglo-Japanese Treaty of Commerce and Navigation* in place between the two nations. It was not until 1939, however, that Roosevelt announced that the U.S. would pull out of this treaty. In 1940, Roosevelt declared a partial embargo of U.S. shipments of oil and metals to Japan.

On December 8, 1940, almost one year to the day prior to the Pearl Harbor bombing, President Roosevelt had a very ominous lunch at the White House with two key people that would set the tone for a future attack on the Japanese. The two people were Henry Morgenthau, Jr., the Secretary of the Treasury, and Dr. Soong Tse-Ven, Nationalist China's Foreign Minister and the man who financed Chennault's Flying Tigers. Morgenthau's notes from the meeting were in writing and clearly stated that Roosevelt had said, "It would be a nice thing if China bombed Japan."

The following day, on December 9th, 1940, Dr. Soong delivered a handwritten note to Morgenthau which included a map of secret air bases in China within flying distance to key Japanese cities. This was one year before the Pearl Harbor attack.

The next day, on December 10th, 1940 at 8:40 AM, Morgenthau met with Secretary of State Cordell Hull and said, according to Hull: "What we have to do, Henry, is to get five-hundred American planes to start from the Aleutian Islands and fly over Japan just once. That will teach them a lesson. If we could only find a way to have them there (SIC) Chinese drop some bombs on Tokyo."

This was the beginning of a continuous series of events which demonstrate the Roosevelt administration's intent to wage war against Japan, all of which led to Pearl Harbor. I will continue to delineate them chronologically.

On December 19th, 1940, Morgenthau, Hull, Secretary of War Henry Stimson and Secretary of the Navy Frank Knox (together, these four were known as the "Plus Four") met with President Roosevelt who expressed maniacal delight at the idea of having ostensibly "Chinese" aircraft bomb Japan. The President directed the "Plus Four" to "work it out."

On December 21st, 1940 at 5:00 PM, Morgenthau conspired with Dr. Soong and Captain Claire Chennault. Chennault presented Morgenthau with a plan to bomb Tokyo and other key Japanese cities employing either (or both) B-17 Flying Fortress Bombers or Lockheed-Hudson Bombers. Incendiary bombs would be deployed to set the Japanese cities aflame and simultaneously save weight so that the bombers could carry more fuel.

DRUGS AND WAR MONGERING

In January of 1941, FDR was informed by the FBI of Lauchlin Currie's operational status as a Resident Foreign Communist Agent. The President's immediate response was to dispatch Currie out of the country to prevent his arrest by sending him on mission to China for coordination with the Chinese Communist Party. That very same month, the United States Naval Security Group broke the Imperial Japanese Naval Codes and the U.S. distributed the I.J.N. Decryption Codex to Britain and all of America's allies.

Once in China, Lauchlin Currie met with Chou En-lai, the Communist Chinese Foreign Minister of Mao Tse-tung's Revolutionary Party. Upon his arrival, FDR also put Currie In-charge of the Prewar Chinese Lend-Lease Program to oversee General Claire Chennault's AVG (American Volunteer Group) of "Flying Tigers", an illegal mercenary wing of American criminals as defined under Sections 956 through 960 of the United States Criminal Code which legally restricts "collusive aggressive activity and forbids conspiracies to injure the property of any foreign state, the hiring or retaining of persons within the United States (from which all of the AVG, aka 'Flying Tigers' sourced from) for enlistment in any foreign military service, and the furnishing of money for any military enterprise against the territory of any foreign state."

Frank Knox (the Secretary of the Navy) had his aide, Captain Frank Beatty, write letters of introduction to commanders of Navy and Marine air bases on behalf of Chennault to visit those bases and recruit men to serve in the employ of CAMCO (Central Aircraft Manufacturing Company), an organization that was used as a cover by the United States Government which took the official position that these men were involved in a "commercial venture without any direct participation by the United States Government." Records of the United States Navy refer to Chennault and his coconspirators in this criminal enterprise by stating, "They realize the necessity for keeping the thing quiet and will take due precautions."

On May 12th, 1941, agent Currie was back in the United Stated and presented FDR with a written war plan titled "Aircraft Requirements For The Chinese Government" which explicitly planned the bombing of Japan by Chennault's Flying Tigers flying out of Tchu-Tchouw, Tzanchouw, and Hengchouw to attack Nagasaki, Osaka-Kobe, and Tokyo, respectively.

This war plan was officially Document JB-355, but it was serialized as both 591 and 691 in documentary status to intentionally confuse future retrieval attempts. The acronym "JB" referred to "Joint Board" which was equivalent to today's Joint-Chiefs of Staff. This meant that this war plan was both reviewed AND recommended by scores

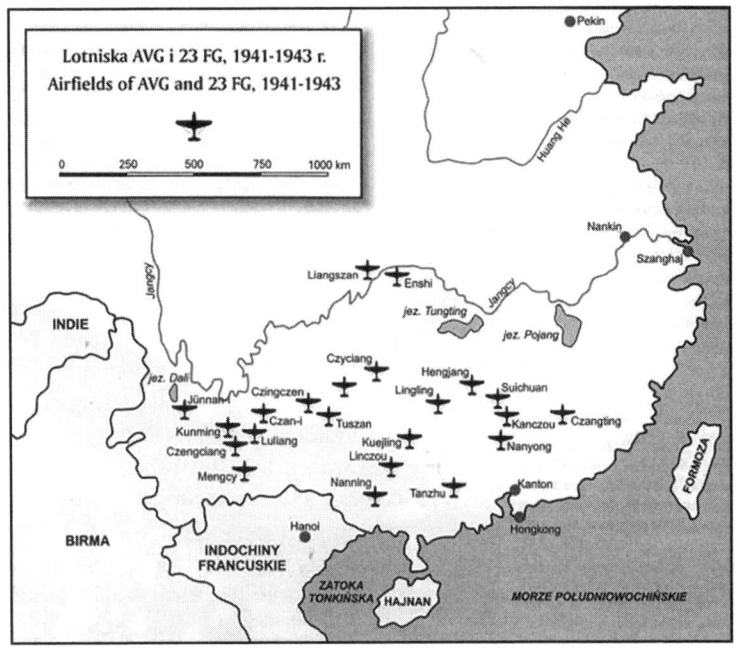

**MAP OF FLYING TIGER AIR FIELDS IN CHINA
(AVG = AMERICAN VOLUNTEER GROUP)**

of responsible contributors including Henry Stimson (the Secretary of War); Frank Knox (Secretary of the Navy — James Forrestal being the First Undersecretary of the Navy at that time who would fill his immediate superior's position upon Frank Knox's death in 1944); as well as General Hap Arnold of the U.S. Army Air-Corps. This is the very act of "conspiring to make war" for which the United States intended to prosecute Axis National Heads of State come 1945—1946. See the document on the next page by the Secretary of the Navy.

In July of 1941, Treasury Secretary Morgenthau advised Franklin Roosevelt that the alliance between the Japanese and Émigré Jewry was an immediate and incalculable threat to American world dominance. The reason for this is that the Japanese had established a buffer state between Manchukuo (Japanese Manchuria) and the Soviet Union. The Japanese had done this in return for their being significantly financed by Jewish interests during the Japano-Soviet Wars of 1937-1939. This buffer state was known as Yevrey and was a Northeast Asian Zion, it being a forerunner to the current state of Israel. As Morgenthau claimed that Jewish interests controlled so much of the U.S. press, broadcast

DRUGS AND WAR MONGERING

> WAR AND NAVY DEPARTMENTS   SECRET
> WASHINGTON
>
> JUL 1 8 1941
>
> S E C R E T
>
> The President;
> The White House.
>
> Dear Mr. President:
>
> At the request of Mr. Lauchlin Currie, Administrative Assistant to The President, The Joint Board has made recommendations for furnishing aircraft to the Chinese Government under the Land-Lease Act. These recommendations are contained in the Joint Planning Committee report of July 9, 1941, J.B. No. 355 (Serial 691), which The Joint Board approved, and which is transmitted herewith for your consideration.
>
> In connection with this matter, may we point out that the accomplishment of The Joint Board's proposals to furnish aircraft equipment to China in accordance with Mr. Currie's Short Term Requirements for China, requires the collaboration of Great Britain in diversions of allocations already made to them; however, it is our belief that the suggested diversions present no insurmountable difficulty nor occasion any great handicap.
>
> We have approved this report and in forwarding it to you, recommend your approval.
>
> Acting Secretary of War
>
> Secretary of the Navy.
>
> 1 Incl.

media, and film industry, Japan was now the only "Colored" country on Earth with direct commercial and intelligence access into "White" markets the world over. In other words, the Japanese affiliation with the Jews could circumvent the boycotts that had been placed upon them by the United States. Accordingly, Morgenthau recommended Restrictive American Immigration Regulations to prevent "Axis-Affiliated Jewish Insurgency into the United States", a recommendation that went into effect in July of 1941 that would literally cost the lives of millions of Jews by prohibiting them access to U.S. Asylum throughout the Holocaust.

On July 22nd, 1941, per direct Presidential authorization, Lauchlin Currie cabled General Chennault to mobilize his criminal mercenary wing, the Flying Tigers, for the falsely-flagged bomber offensive with sixty-six (66) bombers to-be-delivered in-country to China Command. This was criminal as defined by International Law as recognized by the League of Nations, of which the United States was the founder. This would be comparable to either Presidents George W. Bush or Barack Obama having ordered Blackwater (a mercenary corporation in the Mideast) to invade Iran in Iraqi uniforms.

On July 23rd, 1941, this conspiracy was authorized and signed into Executive Order by FDR himself, thereby implicating President Roosevelt as the preeminent War-Criminal of the Twentieth Century.

On July 25th, 1941, five months prior Pearl Harbor, the first 24 Lockheed-Hudson Bombers arrived in China while the U.S. Army Air-Corps was relocating the entire 19th Bombardment Group that was based out of Hickam Field in Hawaii to the Philippines, with the express intent to bomb Taiwan (integral to the Japanese Empire since 1895) to devastate Taiwanese airfields and prevent aerial reinforcements from launching to defend the core population-zones of Japan's major cities.

Finally, on July 26, 1941, and concurrent with his mobilization on other fronts, President Roosevelt cut off the Japanese Empire from all American oil, thus leaving Japan's Imperial Navy with just eighteen months supply of fuel. This was, however, far more than a trade embargo. FDR froze ALL of Japan's AND China's assets in the United States and throughout the U.S. dominated Western Hemisphere (the largest Japanese population in the world outside of Japan, both than and now, is in Brazil), a devastating Act-of-War in and of itself that eviscerated Japan's laboriously accumulated International Monetary Reserves.

By 1941, war all across the globe had congealed the financial systems of all the world's powers into autocratic blocs, thus rendering their currencies incontrovertible. Thus, the Japanese yen was suddenly rendered illiquid; that is, not acceptable for payments outside of the Japanese Empire. The United States stood in the extraordinary position of controlling nearly all of the world's negotiable monetary resources, and it abused this exceptional power to bankrupt Japan with malevolent intent to devastate any Japanese ability to counter-mobilize its own national economy to respond to America's premeditated bombing offensive.

Freezing Chinese assets was intended to block Japanese access to the outside world by reason of their connection with Victor Sasson, the President of the Shanghai Stock Exchange who had helped fund them during their war with Russia. Shanghai represented an international market via their port. Not only did this become inaccessible to the Japanese, the financial strangulation also condemned millions of Chinese to starvation.

In spite of this, Japan had ample liquid assets, including dollars in U.S. banks and gold bars in Tokyo vaults. These were used to pay relatively small international debts incurred by the China War and combatting an illegal insurgency that the Americans were conducting via Chennault's Flying Tigers. Although Japan was not insolvent, FDR's all-powerful financial pen-stroke rendered the entire nation of Japan illiquid. This fiscal freeze isolated Japan economically from the outside world, voiding its monetary assets, both with regard to its sums on-hand or its obtainable currency for the foreseeable future.

This strategic economic offensive was designated as the ABCDE (an acronym for "American, British, Chinese, and Dutch East Indies") Encirclement by the Caucasian imperial powers who instituted it. This included the British and Dutch colonial empires that controlled enormous areas of the Asian-Pacific theater surrounding Japan. These empires enforced parallel economic freezes at FDR's insistence, all of which were fully intended to bring Japan to its knees while American bombardment commenced.

On September 30th, 1941, Roosevelt wrote a secret memorandum to Knox and Stimson declaring that the United States was sending China another 269 P-40 Fighters as well as another 66 bombers to reinforce the one hundred 100 P-40 Fighters already in Burma. The latter were purposely stationed in Burma so that, in the event that the Japanese were to attempt striking the American aerial insurgency at its source-point within the British Imperium, the Empire of Japan would perforce be attacking Pax Britannica itself who would in turn be compelled to engage Britain's multimillion-man Colonial Army of the Raj against Japan. The American buildup of air power was so flagrant that they were confronted by Ambassador Kichisaburo Nomura of Japan who literally stated to the American government that what they were doing at this point was an act of war.

Roosevelt's nefarious plans to bomb Japan were completely uncovered when the Emperor's Kempei-Tai (literally translated as "Thought Police" in Japanese and who served as their version of the Gestapo) broke up the Sorge Spy-Ring on October 18, 1941. Sorge refers to Richard Sorge, a German national who joined the Communist Party, immigrated to the Soviet Union and made his way back to Germany where he would serve as a direct spy for Stalin. While his story is of historical record, his spy-ring and how it precipitated the Pearl Harbor attack is one of the most overlooked and under appreciated pieces of World War II history.

A naturalized German who spoke the language perfectly, it was easy for him to infiltrate the Nazi Party's Diplomatic Corps, whereupon he pulled off a monumental intelligence coup by discovering Germany's planned invasion of the Soviet Union well ahead of time. Stalin, however, thought Sorge to be a fool and rejected the information. Once the actual invasion (Operation Barbarossa) commenced, it was Stalin who looked foolish. Now suspecting that Sorge was a potential double-agent, Stalin ordered him to arrange for his own diplomatic transfer to Tokyo, maintaining his professional cover as a German diplomat while establishing the Sorge Spy-Ring to act as a traceless "Dead-Drop" relay between Lauchlin Currie and Moscow. A dedicated and proven Soviet agent who reported directly to Stalin, Currie was at the core of the United States-Chinese Communist insurgency loop. Stalin wanted Sorge to monitor the progress of Currie's successes in manipulating the United States into attacking Japan. This was key to Stalin because it presumably removed any subsequent threat of the Japanese opening a second front of the Eurasian War in the Soviet Far East.

After this "Dead-Drop" relay was in operation, Lauchlin Currie reported to Stalin, via Richard Sorge, that he had succeeded beyond his wildest dreams, and he credited this as being due to FDR's own personal agenda towards sponsoring global communism. Stalin, however, distrusted Sorge once again, convincing himself that Sorge was a deep-penetration agent leading him astray. Stalin never dared to even imagine that the American people would ever allow an avowedly socialist president to drag them into what could only develop into an unprecedented military debacle in Asia for the sake of furthering the advance of communist revolution both abroad and within their own homeland.*

The reason for Stalin's obstinance in accepting what Sorge either reported or relayed had to do with the dictator's provincial upbringing in the Caucasus Mountains of Georgia. Like America's Ozarks or Appalachians, the ethnicity of this area is steeped in the regional traditions of supernatural lore. While the concept of the "Evil Eye" is readily accepted and well-known, the Caucasiatic "Eye of Joy" is less understood. It conveys the belief that some things are just too good to be true, and are therefore to be rejected.

Accordingly, Stalin immediately ordered Richard Sorge's wife to be executed as he was now convinced that Sorge would defect to the Axis and that this was his only opportunity for reprisal. Unknowingly, Stalin's superstitious hillbilly paranoia which led him to suspect Sorge would lead directly to the Japanese preemptive surgical strike at Pearl Harbor.

---

*It is a point of fact that the same American people who elected and idolized the President who did so are still in denial of these matters to this very day.

The Japanese had become suspicious of Sorge because of the increasing number of encrypted radio messages he was sending, and it was not just the number of them. His coded transmissions appeared as gibberish, and it became obvious to the Japanese that Sorge was encrypting his messages with unbreakable "one-time pads", a technique that the Soviet intelligence agencies always used.*

Suspecting Sorge of being in league with the Russians, the Kempei-Tai (the Japanese equivalent of the Gestapo) arrested him in Tokyo. Serving Hirohito directly, the Imperial Japanese Secret State Military Police tortured Sorge mercilessly. When he proved himself willing to die rather than be broken, the Kempei-Tai presented him with irrefutable evidence that Stalin had ordered the murder of his wife. It was this fact which finally convinced him to confess to his captors FDR's extant plans to bomb Tokyo and other major urban population centers.

Upon learning of Richard Sorge's apprehension, the FDR administration assumed that the Soviet Union's master spy would break under pressure and expose their own active conspiracy to escalate the contemporary undeclared East-Asian conflict to unprecedented levels. Their immediate response was to encourage America's national newspaper of the day (comparable to today's newspaper *USA TODAY*), *The United States News*, to "go public" and they did, carrying a two-page story on October 31st entitled, *Bomber Planes to Japan: Flying Time From Strategic Points*. This was an absolute must for the administration because, if they did not reveal the logistical facts of all these airplanes, they would have been badly embarrassed if the Japanese revealed it to the world first. The story included an illustration depicting bases in Chunking, Guam, Dutch Harbor in Alaska, Cavite City in the Philippines, Singapore, and Hong Kong, the latter two of which the British were allowing the Americans to use; whereas, the Soviets were permitting the Americans to use Vladivostok. The newspaper announced that American bombers would be launched from these areas to bomb Japan.

On November 1, 1941, with intent to attack Japan, the U.S. Army established the Military Intelligence School (MIS) inside El Presidio of San Francisco's Crissy Field Building #640. Fifty-eight U.S. Citizens of Japanese ethnicity were placed under the command of Lieutenant-Colonel John Weckerling to study Japanese under the tutelage of one of the only Caucasians in the United States fluent with the Japanese language. At that time, most Japanese-Americans were so well assimilated that there very few of them who spoke Japanese, and they had to

---

*A "one-time pad, the acronym for which is "OTP", is an encryption technique that requires the use of a one-time pre-shared key. It cannot be cracked without the pre-shared key. The key consists of plain text paired with a random secret key that is only used for one message.

have Caucasian tutors. These Japanese-Americans were intended to be deployed as saboteurs and spies against the Japanese Empire.

On November 15th, 1941, United States Army Chief of Staff, General George Catlett Marshall, conducted a secret briefing before the American press and this included the *New York Times*, the *New York Tribune*, *Newsweek*, and the Associated Press. To this assorted network of embedded reporters, Marshall asserted to them that bombing operations would be initiated by the United States against Japan within the first ten days of December.

On November 21st, 1941, the United States Army Air Corps Strategists commenced selection of civilian population centers within Japan to be attacked by heavy bombers. On the very same day, the 2nd AVG (Second American Volunteer Group) – the men that would man and service the so-called "Chinese Bombers" – set course for China and sailed from San Francisco. Sitting on the tarmac in Burbank, California were Lockheed-Hudson bombers waiting to be flown across the Pacific. The United States was fully mobilized to carry out total war against the Japanese Empire.

Immediately after learning of Franklin Roosevelt's plan to completely subject their Empire, the Japanese prepared by spending the following month mobilizing a simultaneously coordinated series of preemptive surgical strikes against ALL American forward strike-bases where U.S. strategic heavy bombers were amassing to attack the Japanese civilian population. All of this amounted to an entirely justified and purely defensive military prophylactic strategy which was, of course, inclusive of Pearl Harbor.

Although the U.S. knew how to break Japanese intelligence codes and therefore knew of the Pearl Harbor attack before it happened, the Office of War Information has forever since propagandized this event as a "surprise attack" to serve as a rallying cry for a war that was not really necessary from an American perspective but was motivated by FDR's Rainbow War Plan in an effort to maintain their lucrative opium trade.

CHAPTER EIGHT

PEARL HARBOR

The Americans had broken the Japanese diplomatic code as early as 1937, but the most important thing was the fact that the Americans, once they found out the Japanese were going to attack, decided that they were going to hold an ambush. The code breakers knew the Japanese attack was going to come in three waves. The first wave was when the Japanese would initialize the attack and bomb the ships in the harbor. The second wave targeted the airfields. As the Americans knew the codes, they were fully expecting these attacks. The third wave, however, never happened and that completely undermined the Americans' plan for an ambush. I will explain.

What is generally believed about the Pearl Harbor attack was that most of the fleet was in the harbor like sitting ducks, save for the Navy's aircraft carriers. What is not realized is that the ships at Pearl Harbor were all very old World War I dreadnoughts that were not about to survive another war. The Americans were planning on scuttling them anyway, and they figured that the Japanese would be doing them a favor by sinking them. For that reason, and quite oddly, they were literally chained right next to each other at anchor. The Americans also crudely modified an old battleship, the *USS Utah*, to appear to be a carrier. This was a pure decoy to draw the Japanese in. As soon as the Japanese would attack at dawn, the Americans were planning to come in like the cavalry in a cowboy movie and they would rescue the men under attack. Although it was a deliberate set-up, the Americans did not expect to get caught with their pants down; and unable to respond to a third attack that never materialized, this is exactly what happened.

The name of the scheduled American operation was "At Dawn We Fought", but it turned out to be more like "At Dawn We Slept", the latter actually being used as a title for a book by Gordon W. Prange that exposed severe American incompetence on December 7th. Unfortunately, such incompetence is incorrectly assigned to Admiral Kimmel and General Short, both of whom were chosen to serve as scapegoats.

Fully expecting an attack, the Americans had pulled out the vessels they cared about: submarines and carriers. The ships left behind at Pearl were neither fueled nor loaded with ammunition. This was also the case for most of the planes that were bombed. Those were ready for shipment to China. There was, however, another sinister aspect to the American plan, and that had to do with the sailors who were manning this faux fleet.

The only men at Pearl Harbor — and you can check the names on the roster — were all young swabbies between sixteen to eighteen years old; and all of them were given a big drunken party the night before the Pearl Harbor attack because the high command knew they were going to die. All of the old veteran sailors with indispensable combat experience — men like my father and other veterans who had fought in the Chinese theater or World War I — they were still needed. You did not find any of them ashore at Pearl Harbor that fateful day.

It was the task of the aircraft carriers to perform evasive maneuvers in order to carry out the ambush against the Japanese fleet. More specifically, the *USS Lexington* had just left Pearl Harbor to deploy 18 Vought SB2U Vindicator dive-bombers on Midway Island to attack the approaching Japanese task-force from the north, thus trapping it in a pincer assault. The *USS Saratoga* was being held in reserve in San Diego to conserve force assets in case the anticipated Battle of Oahu resulted in any American carriers going down.

Although Admiral Kimmel and General Short received the blame, the failure of "At Dawn We Fought" focuses on Admiral Halsey, the commander of the *USS Enterprise*, who took his ship and delivered 12 Grumman F4F Wildcats to Wake Island where a Japanese attack was anticipated. The Wildcats were to box in the Japanese Pearl Harbor Task-force from the rear. When the code breakers informed him that the Japanese had changed their plans and were going to focus their attack on Pearl Harbor, Halsey began his return to Hawaii. Told by Admiral Nimitz to use his common sense, Halsey issued his now famous "Battle Order Number One," the first item reading, "The *Enterprise* is now operating under war conditions."

When Halsey ordered his scout planes to "sink any shipping sighted, shoot down any plane encountered," his operations officer challenged this order and Halsey replied, "I'll take [responsibility]. If anything gets in the way, we'll shoot first and argue afterwards." His intentions were to bomb anything on the sea and shoot down anything in the sky.

When the *Enterprise* returned from Wake Island, she remained hundreds of miles off the coast of the southwest of Hawaii and was ordered to maintain radio silence under Admiral Halsey while lurking to ambush the Japanese attack. What ensued under his command, however, still lives in infamy that, while not completely unrecognized, is mostly overlooked so as to give a complete distortion of what really happened on December 7, 1941.

PEARL HARBOR

At 06:15 in the morning, exactly the same moment when the Japanese launched their first attack wave, the *USS Enterprise* under Admiral Halsey launched Scout Squadron 6 in a probing attack consisting of 18 Douglas SBD Dauntless dive-bombers. They actually arrived at Pearl Harbor at the same time as the Japanese bombers and engaged them in aerial combat. As this engagement had been radioed back to Halsey, there is no way anyone could say he was ignorant of the attack.

Scout Squadron 6 was forced to scatter when a Japanese Zero rammed Ensign J.H.L. Vogt's bomber and disintegrated it in mid-air. This amounted to the first Japanese Kamikaze attack of the Japanese-American war and this denied Halsey extended reconnaissance. Although he had a fully loaded and operational carrier waiting to come in, Halsey chose to sit out the next wave of attack and allowed thousands of American servicemen to die. He was waiting for a third wave that never happened.

As the Japanese bombed Hickam Army Air Field, Captain Brook Allan's quick thinking saved his B17 from destruction when he taxied it away from the flight line and managed to take off, rising alone into the afternoon sky with orders to search to the southwest. Finding Admiral Halsey's huge carrier, they opened fire on him and tried to shoot him down. This shows you how bad the situation was.

It was not until 17:00 (5 PM), which was hours after the initial Japanese wave, that the *Enterprise* began to launch a viable attack force, and this included eighteen Douglas TBD Devastator torpedo-bombers and six of the remaining SBD Dauntless dive-bombers from scouting squadron 6 which had originally tangled with the Japanese on the first wave. Fitted with smoke generators to mask the lumbering torpedo planes, they also escorted 6 Wildcats. After launching to attack the Japanese task-force, they soon discovered that the Japanese had left. As Admiral Nagumo recognized that the Americans had set up Pearl Harbor as a trap, he withdrew the third wave and it never happened. The Americans were caught with their pants down.

Mitsuo Fuchida, the Japanese pilot who led that first wave attack, had spotted something very odd with regard to the American "fleet". He was quoted as saying, "I've seen German warships assembled in Kiel Harbor. I've seen French battleships in Brest, and I've seen our own warships assembled for review before the Emperor; but I've never seen ships — even in the deepest peacetime — anchored at a distance of less than five hundred to a thousand yards from each other."

Further, upon seeing all those battleships berthed side by side, Fuchida said the picture was just hard to comprehend. And so, the Japanese realized something very strange was going on. The *USS Utah* was a converted battleship made to look like a carrier. When they sank it, they realized that it was a dummy target ship and the Japanese pulled out. And, of course, there were no American submarines or carriers there; so, by not accommodating the Americans with a third wave of attack, the American high command were made to look like idiots, and I think they were more angry about that than the attack itself. In fact, it was far worse than you might imagine.

When the Wildcats and the other planes Halsey had sortied reported that the Japanese had left and evidently abandoned their plans for a third wave, they proceeded to return to the *Enterprise*. This, however, presented a huge problem since Halsey refused to turn on the landing lights because he was afraid the Japanese might be circling around to destroy him now that he had an empty carrier. Therefore, the Wildcats and all of the planes tried to land at Pearl Harbor but because Halsey refused to contact Pearl Harbor by radio (again, for fear of transmission intercepts leading to him being located by the Japanese), the aircraft were shot at by their own men on the ground. Everything opened up. The whole night sky was filled with tracers, and the only plane that survived was flown by Wildcat pilot Ensign James Daniels. By turning on his brights — his landing lights — he blinded the anti-aircraft gunners and was able to land, but everyone else met with great misfortune. When Ensign David Flynn's plane ran out of gas, he parachuted into a cane field. Lieutenant Eric Allen Jr. was not so fortunate as he was shot out from under his parachute. Ensign Herbert Menges and Lieutenant Francis Hebel were shot down in their planes. All of these young men died as a result of being shot down by American "friendly fire" because Halsey refused to radio Pearl Harbor to tell them these boys were trying to land.

The most chilling aspect of what I read while at the Presidio concerned the Salvage Divers (Deep Sea Divers) and Underwater Demolition Team (UDT) "frogmen" (wet-suit divers) who went down to investigate all the ships. Ostensibly, their purpose was to make sure that there were not explosive levels of fuel that were leaking. They were doing what they could with the fuel leaks and also clearing what wreckage they could to keep it from blocking the harbor. The dive teams were very active, and they heard the boys trapped inside the *Arizona* and all these other sunken ships like the *Utah*. They were tapping for help in Morse code for five and even up to thirty days after the attack, but the divers were given orders by the high command to do absolutely nothing to rescue them. What can we make of the inhumanity of ignoring these boys and swearing their would-be rescuers to secrecy, only to leave them to die in the darkness and the damp?

If the men had no access to any kind of facilities or food or any of their rations, we can assume they would have had to have resorted to cannibalism at some point. We can only speculate as to exactly why the high command issued that life-choking order, but based upon what I read, it was obvious there were preset demolition charges in all of those ships. Inexplicable explosions were felt on each of the ships throughout the Japanese attack, and while the survivors' accounts of these phenomena were later attributed to collateral fires setting off ammunition, boilers, etc., the circumstances speak for themselves. Were the Japanese to fail to take these ships down for any reason, the Americans were prepared to take them down on their own. That is why they did not want these boys rescued. If the salvage men had attempted to cut them out, many might have drowned, but there was the distinct possibility that at least a few of them might have made it out. Had any witnesses survived, they might have told the true tale of what was going on. Bear in mind, these were young sailors who knew very little about the ways of war; and in terms of needed manpower, they were easily expendable from a strict military point of view.

For their part, the Japanese certainly had an opportunity to destroy the fuel or oil storage tank farms at Pearl Harbor, but they did not do so because they were under strict orders not to cause undue casualties among the civilian population. Although there definitely were civilian casualties, the Pearl Harbor attack was an extremely precise surgical military strike on the part of the Japanese. In light of the provocations of war that had come so repeatedly to the Japanese up until that point, what happened cannot be defined in any way as a terrorist attack. What the Americans had done to their own sailors, however, was something far worse.

THE ROSWELL DECEPTION

CHAPTER NINE

The Chrysanthemum Throne

The Chrysanthemum Throne is the name for the Imperial Throne of Japan, recognized as the oldest continuing monarchy in the world. According to legend, the Empire of Japan was founded in 660 BC by Emperor Jimmu and the subsequent monarchs of the Chrysanthemum Crest, including Hirohito and his family, are direct descendants of him. The Emperor serves as the high priest of the ancestral religion known as Shinto, and prior to and during World War II, the Constitution recognized the Emperor as being of divine origin.

Although he was the longest-lived and longest-reigning historical Japanese emperor and the longest-reigning monarch in the world at the end of his life, the pivotal role that Emperor Hirohito played before, during and after World War II was formidable; and besides being unappreciated and unacknowledged, it is largely unknown.

The biographical history of Emperor Hirohito is so complex that it would require at least an entire volume to give a comprehensive assessment of the man and his role in history. Even then, so much of the complex dynamics influencing the psychology of his culture and his persona would tend to dumbfound the typical Western mind.

His role as high priest of the Shinto religion can be better understood to the Western mind as being like the Pope. In this capacity, he received a tithe of ten percent from all of the subjects in his Empire. Besides being a scientist, he was both their secular and religious leader.

An example of how the Western mind can convolute Japanese culture, or any culture for that matter, is when academics attempt to make an assessment of the religious nature of the population. They might divide it into different percentages of Buddhists, Christians, Shintoists and so on. This will only confuse and distort the issue, not even recognizing the actual truth that they cannot comprehend.

While Shinto is recognized as the indigenous religion of the Japanese with Buddhism being imported in the Sixth Century, it is not that simple. Although Siddhartha Gautama is popularly credited with originating what the world knows as Buddhism, it did not begin with him. The same can be said for Christianity and even Judaism, at least as far as these faiths have been understood in the West.

The fact that "Christianity" was reported to have been outlawed by the Shogunate confuses such issues even more. Buddhism, Christianity and Judaism have ancient roots in Japan that are completely obscured by modern "scholarship", but they are a separate and very involved subject that we will not address here. What is important in the current context is that the culture of Japan has been embracive of all these threads and this applies to the Showa Emperor as well, his ecclesiastical reach extending into all of these different faiths.

In the context of this narrative, I want to convey that the overwhelming majority of Westerners have no understanding that Apocalyptic elements are intrinsic to Buddhism, and all of it applies to the Emperor's role as "Pope" to the Japanese culture.

According to a view of Cosmic History widely held in almost all Buddhist cultures, the period following the death of Siddhartha Gautama is divisible into the Three Ages of Buddhism and, potentially, a Fourth (Posthistorical) Age of Enlightenment.

The first two Ages are the Age of Right Dharma ("Shobo" or the "Age of the True Law") and the Age of Semblance Dharma ("Zobo" or the "Age of the Facsimile Law"), followed by the Age of Dharma Decline ("Mappo" pronounced "Mahp-Poh", the "Latter Day of the Law") – the degenerate Third Age when people become lawless and forget the Buddha's teachings.

Traditionally, this Last Age was supposed to begin two millennia after Siddhartha's passing and would itself last for ten thousand years. Temporally (not "temporarily"), this Age of Anxiety in Japanese History commences from the Mid-Eleventh Century. In those days, it was specifically feared that society had entered or was about to enter Mappo – the Latter Day of The Law. By Japanese calculations, the Third Age of Mappo commenced in AD 1055 of the Common Era. This anxiety over Mappo shaped many aspects of Medieval Japanese culture and served as a catalyst for the growth and renewal of Buddhism.

By the time of the Twelfth Century, chaos and change overtook the seven thousand islands of the Greater Japanese Empire. Governmental organization changed dramatically, and the economic structure began to shift. There were innumerable battles of varying sizes as well as a number of devastating natural disasters. There was little or no governmental control over much of the country for much of the time. All of this could not help but have an effect upon the religious climate of the nation. The climate of the Age was viewed through Apocalyptic Buddhism which provided context in which the chaos of the century could be interpreted.

As this interpretation of Buddhism was hardwired into their culture, Japanese "Mappo-anxiety" continues to influence events through to this very day. During this degenerate Third Age, it is believed that people are unable to attain enlightenment through the Word of Bodhi Shakyamuni (Siddhartha Gautama) and that society becomes evermore morally corrupt.

In Buddhistic thought, the teachings of the Buddha are still correct during the Age of Dharma Decline, but people are no longer capable of following them. Even so, Buddhistic temporal cosmology assumes a cyclical pattern of Ages; and when the current Buddha's teachings inevitably fall into disregard, a new Buddha will be born to ensure the continuity of Buddhism. A new period, in which the true faith will flower again, will be ushered in sometime in the future by the Bodhisattva ("Buddha-To-Be") Maitreya ("Miroku" in the Japanese) into the Era of "Shoho", a Time of Peace in which society will be in harmony with Buddha's precepts.

So it is that when Hirohito assumed the Chrysanthemum Throne in 1926, he was entering a theater of circumstances that were steeped in a tradition of apocalyptic psychology within the context of an ancient faith that he was the steward of. The fact that he was soon confronted with a war in China, a conflict that had been brewing since long before he came into power, is only one major thread in a degenerate Third Age of lawlessness where Buddha's teachings were neither recognized nor followed.

It is crucial to keep all of this in mind as you read the intense complexities of the Emperor's actions. From certain perspectives, he can be viewed as a notorious war criminal who had looted all of Asia for his own benefit. From other perspectives, he can be viewed as a genius warrior who had a far seeing insight into the minds and madness of his enemies and was able to beat them at their own game so as to secure peace and prosperity for the civilization that he was the spiritual and temporal guardian of. While it can be argued that there is no ultimate justification for immoral behavior such as killing human beings, it is eye-opening, if you want to begin to understand the true nature and complexity of Hirohito's life, to recognize the role he was born into.

At the time of Hirohito's ascension to the throne, Japan was the only Asian nation other than Thailand that was not colonized or serving the interests of Western powers. The British had "crucified" India by planting hedgerows in a cross-like pattern across the entire subcontinent of India, thus dividing it into four quarters. This made it impossible for the country to unite against British influence and led to the country being completely colonized. The French had taken over Indochina, the Spanish the Philippines, and the Portuguese and Dutch also had their colonies. There was also the earlier mentioned case of Durham White Stevens, the career diplomat who was serving the interests of the Union Pacific railroad and wanted to expedite laying the rail-link between Japan and the rest of Asia via the mainland under control of American stockholders. Japan stood alone.

In the case of the American invasion of the Philippines, President McKinley paid Spain $20 million dollars for those islands and declared a period of "benevolent assimilation". For McKinley and America, this meant a military base and access to Asian markets. For the Philippines, it mean colonization and exploitation of their citizens and resources.

There is no question that the roots of Japan's impetus towards war at the time of Hirohito's coronation go back to the invasion of Commodore Perry. That incident created a completely different political psychology in Japanese politics. It became very clear to the Japanese that if they did not embrace the industrial revolution that was taking place across the world that they would be overrun by it.

The main point here is that Japan had to be aggressive in order to protect its own sovereignty. To do otherwise would result in being overrun by Western colonial interests. In embracing the industrial revolution, Japan had thousands of years of cultural history to rely on which it would eventually use to subdue its adversaries and secure its position as the leading power in Asia. All of this was orchestrated by Emperor Hirohito, but the process took decades to unfold, and it is remarkable that any world leader could persevere through such trying circumstances, let alone succeed in his objectives which, against great odds, he eventually did.

This was the perilous position Hirohito found himself in upon assuming the role of Emperor and facing a war with China, the cause of which involves a long and complex history going back to the Mongul invasion of Kublai Khan. There was also another major factor that was even more menacing than colonization. That was the agenda proliferated by Jack London and the weapon of mass destruction that the Americans had deployed to end World War I. The seeds of war were in his mind, and the Emperor would have to make the best of them.

Prior to his ascension to the throne, Hirohito was invited by the British Crown to visit England whereupon he became a Knight of the Royal Garter, the most senior order of knighthood in the British honors system. During his visit to England, he realized that they had a Royal Air Force. At that time, every modern industrial nation in the world had an air force except for two: Japan and the United States. While the U.S. had the Army Air Corps and Naval Aviation, there was no separate branch of service for aircraft. When Hirohito became Emperor, he was advised by Admiral Takijirō Ōnishi that if he were to start an air force, he should keep it secret from the Americans, making it covert and start it as a black budget program that the Americans would not find out about until it was too late. The Emperor followed his advice to the letter. Additionally, he would concoct a plan to use his secret air force to carry biocidal weapons of mass destruction to both China and the United States.

As a marine biologist, Emperor Hirohito was the first trained scientist in charge of a nation-state and he took advantage of his studies so as to use his enemy's own weapons against them, and here I am referring to biological warfare. Understanding the nature of wind currents and how they had affected the ecosystem and history of Japan over many centuries, Hirohito studied what is known as the Kuroshio, one of the three largest of the world's ocean currents. Running up the eastern coast of Japan, it continues its movement northward until eventually crossing the Pacific Ocean to the West Coast of the United States before running southward and eventually back across the ocean to Japan. It is the equivalent of what is known as the jet stream and runs clockwise in only one direction.

Long fascinating humans, early fishermen and explorers took note of the Kuroshio and other currents because these either sped up their voyages or got them lost. Early Chinese mariners called the Kuroshio Current "Wei-Lu" or "the current to a world from which no man has ever returned". The Japanese named it "Kuroshio" or "black current" for its dark cobalt blue waters. Physical oceanographer Steven Jayne has said, "The Kuroshio is the strongest current in the Pacific Ocean, and is also one of the most intense air-sea heat exchange regions on the globe. It influences climate as far as North America."

At the time Hirohito studied the Kuroshio, Western science had no idea of this potent force of nature, and their first encounter with it was when their bombers hit up against it during the war. The current itself was a weapon of war and could be used to send lighter-than-air balloon-bombs and dirigibles with biological weapons of mass destruction. This was all the Emperor's brainchild, and he was harnessing nature, the hallmark of the Shinto religion.

It is important to recognize that Hirohito was not only the first head of state who was an actual scientist, he had a whole country at his disposal. To put this into perspective for you, the average scientist in the world is impotent and has no social power. He has no ability to control the direction and outcome of his work other than to do the research that he is funded to do. Hirohito was the only scientist in the history of all Mankind who was ever a head of state. Since that time, there have only been two others, both of them women: Margaret Thatcher of the U.K., a physical chemist, and the other one is Angela Merkel, a research chemist who at this time of writing runs Germany as Prime Minister. Hirohito was the only head of state to be so, and as Emperor, he was worshiped as a god. Like a Pope receives a tithing from each church, everyone in Japan gave him ten percent of their salaries, the result being that when he overran Asia and took all of the national treasuries of the Philippines, Korea, and all of the other kingdoms and empire states, he possessed the equivalent of a hundred billion dollars at the end of World War II. Deposited in Swiss bank accounts, this hundred billion dollars were World War II era dollars, an amount which would add up to over a trillion dollars in today's funds. The Americans never touched it and never could. Able to fund anything he wanted, Hirohito was the only scientist who was a head of state with an unlimited budget. Literally designated as divine by the constitution, Hirohito used his powers to save himself and his country and was actually able to forge a prosperity for his people that is unprecedented in modern times. Hirohito was the most influential man in modern history, and he was a revolutionary pioneer in the field of biological warfare as well as a master strategist in its applications. The world, and particularly America, is fortunate that, unlike Woodrow Wilson, he demonstrated considerable restraint when it came to unleashing his weapons as he was fully capable of destroying the United States.

Japan's war with China not only required necessities beyond the ordinary military means of the time, it was also an opportunity for the Emperor to field test his master strategy against America. To understand this, it is first necessary to take stock of the actual geography of China. A country that is the size of the United States on a map, most of China is an arid and unlivable terrain of rocks, sand, dirt and ice. In the 1930s, one out of every four people on Earth was Chinese; and further, that 25% of the world's population was only able to live in 25% of their own nation. The Japanese, whose population was one hundred million, were facing a war against a billion people. The only possible weapons they could use against such massive odds were weapons of mass destruction. It was not, however, going to be simple to deliver them.

As was said previously, China was a nation of warlords living in isolated regions surrounded by hills and was accessible only by rivers. When it comes to military strategy, rivers are one of the most formidable obstacles to overcome when waging war. When Napoleon invaded Russia, he lost 20,000 men to the first river he crossed, and that is the reason he lost the war in Russia. He did not lose it to the Russians but rather to the river. That is how bad it can be for an army. In China, there are terraces. Every single bit of land is used that is available for agriculture. They use their ponds to breed and raise fish for the Chinese

THE CHRYSANTHEMUM THRONE

THE ABOVE MAP INDICATES THAT THE JAPANESE OCCUPIED AT LEAST 25% OF THE ENTIRETY OF CHINA

people to eat. An additional obstruction was that all Chinese population centers from the warlord era were fortified with enormous walls with only very small roads and walkways. You cannot possibly get a modern mechanized army across such territory.

The Emperor's strategy, which proved itself to be ingenious, was to perfect biocidal weapons, design and construct unconventional aircraft that were capable of delivering them, and supplement these efforts with other armaments. Inheriting the Chrysanthemum Throne under such circumstances could be viewed as a nightmare scenario without much choice, unless one wanted to submit to the West. As has been already stated here and will be further elaborated in the rest of this book, the actual history of Hirohito's legacy is so much different than what the Office of War Information and their various successor organizations have sold the world.

CHAPTER TEN

THE CHINESE WAR

Emperor Hirohito secured his biological weaponry through Hideyo Noguchi, a bacteriologist who achieved great notoriety in both Japan and the United States. Noguchi's work was sponsored for twenty-five years by the Rockefeller Institute for Medical Research in New York City where he claimed to have isolated over half a dozen diseases, including trachoma, syphilis, and yellow fever. His research was subsequently highly criticized and found faulty, but this is because he was deliberately lying during his tenure in America so he could have full access to American biological labs. He was actually a bio-scientific spy who cultivated microorganisms that had never before been grown in the test tube. His greatest contribution to Japan's war effort was that he, before dying of yellow fever himself, brought yellow fever home to the Emperor so that it could be weaponized. Hirohito not only weaponized it, he perfected it to such a degree that he was potentially able to exterminate millions.

Emperor Hirohito's most immediate challenge was in ratcheting down the lethality of the yellow fever virus. The model he used was the so-called Spanish Flu (American Army-Navy Flu) which had killed 100 million people worldwide and ten percent of its victims. It edged out the black plague in terms of its lethality. He had actually ratcheted down the lethality of his yellow fever concoction to the level of the Spanish Flu, thereby facilitating its contractibility .

Emperor Hirohito had been known to test these weapons, and the Americans knew he was battle testing these weapons on the fields of China. The battle of Hongkau was a four month campaign where 230,000 Japanese defeated over a million Chinese. The only way they could beat these odds was that over 60% of the Chinese were rendered deathly ill from biocidal weapons that had been placed into porcelain shell pieces that were launched by artillery and exploded upon the population. The spread, from a local perspective, was all the more effective due to the isolation of the Chinese, the population of which did not normally venture more than ten miles away from their homes during their entire lives. In this manner, the biological weapons of Japan could be tested on China and easily contained.

To test these weapons, the Emperor enlisted the services of one of the most overlooked characters in Japanese history, Shirō Ishii, a microbiologist and army medical officer who served as the director of the notorious Unit 731, a department of the Imperial Japanese Army that was dedicated to biological warfare. Directly under the orders of the Emperor, they eventually perfected their weapons of mass destruction by testing them on prisoners of war. First, it was the Chinese, but as the war developed, they took photographs of Americans dying along with Russians, Australians, and British prisoners of war. The reason for this is that, if they did not share such photographic evidence, the Americans would not believe the germs Hirohito created could actually kill white people. The Americans were actually that myopic and regimented in their thinking as they considered themselves to be a superior race and significantly different biologically.

SHIRŌ ISHI

While Shirō Ishii was recognized as a war criminal, he would play an absolutely crucial role in getting the Americans to capitulate to Hirohito's stipulations at the end of the war. Although reportedly arrested in 1946 by the Allies, he was granted immunity from war crimes under the guise of providing information that was "absolutely invaluable" with regard to biological weapons. He later advised the U.S. Army on

UNIT 731 ON THE CHINESE MAINLAND

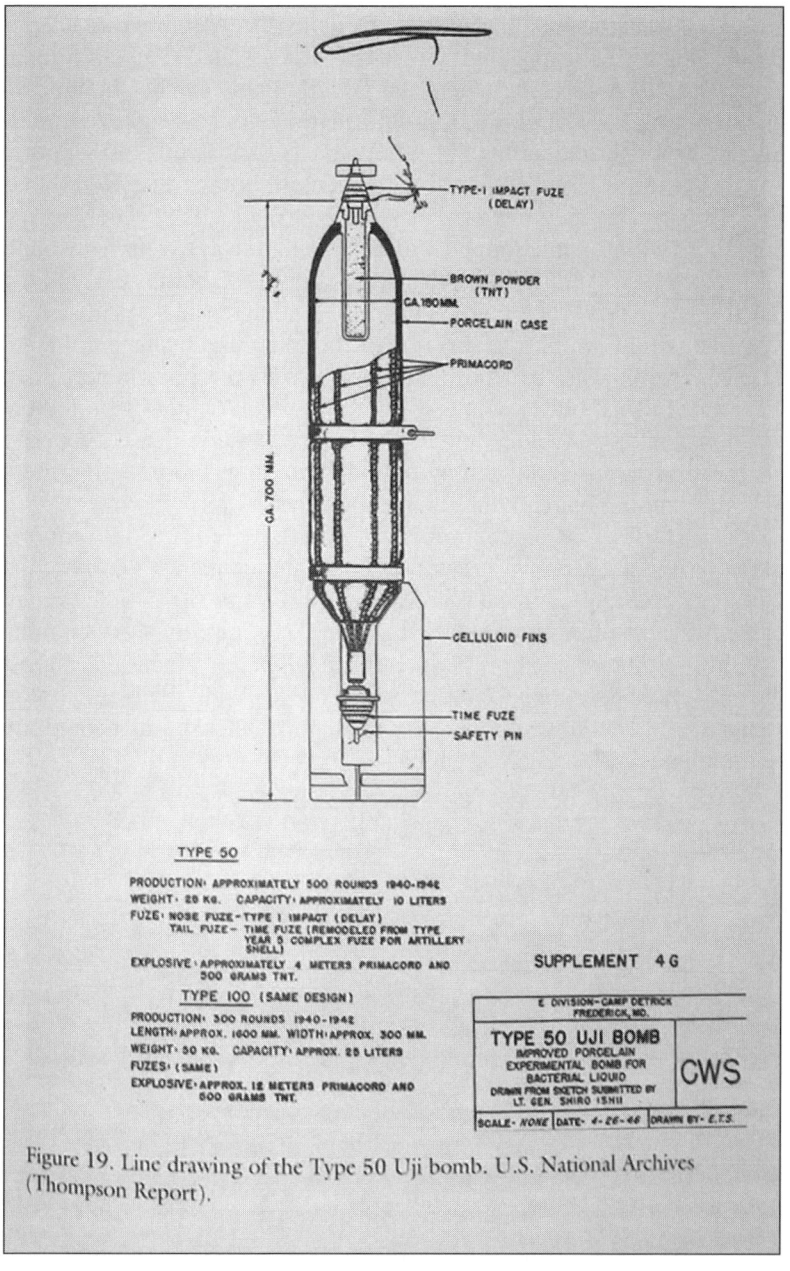

Figure 19. Line drawing of the Type 50 Uji bomb. U.S. National Archives (Thompson Report).

PORCELAIN BOMB DIAGRAM DEVELOPED BY SHIRŌ ISHI

biological weapons at Fort Detrick. Douglas MacArthur wrote a letter concerning Shirō Ishii on May 6, 1947 which stated that "additional data, possibly some statements from Ishii probably can be obtained by informing Japanese involved that information will be retained in intelligence channels and will not be employed as 'War Crimes' evidence."

It was Ishii that developed the porcelain bombs that were burst over China. Dropped from the sky and bearing loads of biocides, the porcelain bombs were dropped with a parachute so as to retard the burst of the biocides. They learned to always drop these bombs early in the morning as that is when there is the least amount of heat to cause friction that would inhibit the survival of the microorganisms employed. The end result was that they perfected this to the point where they were able to wipe out the Chinese Fifth and Sixth Armies by biological weapons. These biocides were so vicious and toxic that they would eat away at clothing and skin with bodies fuming, soon melting them down to nothing but calcium after a few days.

Due to the isolated geography of China, they could not only confine the experiments to specific areas, they could also contain any communist uprisings and any Chinese resistance by breaking the country down piecemeal and ultimately controlling China. It is important to recognize that, at the time of the bombings of Hiroshima and Nagasaki, almost the entirety of the Japanese military was on the Asian mainland, not in Japan itself. The Japanese were very much in control of the geography of mainland Asia, including China.

JAPANESE TROOPS WEARING BIOLOGICAL PROTECTIVE EQUIPMENT

ABOVE IS A PHOTOGRAPH TAKEN BY MY MOTHER OF CHINESE KILLED AS A RESULT OF THE AFOREMENTIONED BIOLOGICAL WEAPONS.

As effective as the biocides were unto themselves, they required an efficient means of deployment, and this is where the dirigibles came into play. As was previously alluded to, these dirigibles had been developed throughout the China conflict because China has an environment covered with ravines, rivers and a very broken landscape. You cannot wage a modern mechanized war in China like you can in Europe. The Japanese developed super dirigibles not only to disperse biocidal weapons but to carry tankettes which were very small tanks that could be lowered into the middle of walled cities in China.

The idea for dirigibles was brought to Hirohito's attention by Admiral Takijirō Ōnishi, the same man who would later develop the Kamikaze suicide units, officially known as the Tokubetsu Kōgekitai ("Special Attack Unit"). It was Admiral Ōnishi who qualified these pilots and turned them into a warrior cult. It was also Ōnishi who approached Hirohito and informed him that they would need dirigibles as opposed to planes or anything else in order to drop tanks into the middle of these Chinese cities. When it came to dirigibles, the Japanese

ABOVE ARE THE RAPIDLY DECAYED REMAINS OF BODIES MELTED DOWN BY REASON OF BIOCIDAL WEAPONS.

TAKIJIRŌ ŌNISHI

benefited greatly from the learning curve of both the Germans and the Americans, the latter being particularly disastrous.

Developed by Count Ferdinand von Zeppelin, the German dirigibles were a huge factor in World War I. Greatly benefiting from at least twenty-two years of experience by the Germans, the Japanese developed silken silver jump suits for pressure protection which were worn by the small Yakuza who operated the craft. This, in conjunction with the fact that these tiny Asians shaved all their body hair to avoid causing burns from electrostatic discharges, prompted many witnesses (outside of those informed) to conclude that the operators were indeed extraterrestrials. See Appendix A for an extensive account of how the German program led to the development of these pressurized suits.

Right after World War I, it was the Americans who developed the concept of a flying aircraft carrier. The U.S. Navy had developed three flying aircraft carriers: the Macon, the Akron, and the Shenandoah. These were gas-inflated dirigibles with a rigid structure of metallic frames. At the same time, they had a catapult system under which a biplane could hook and unhook themselves. As many as five biplanes or "parasite fighters" could be attached to such a carrier. Instead of having their carriers housed on the coast, they sought invulnerability by erecting large hangars deep within the continental United States. This is why you have huge Naval air bases out in the middle of Kansas that are landlocked. They are just huge dirigible hangars. They could float these dirigibles over the Pacific Ocean and deploy the parasite fighters to either attack ships or to defend the dirigible itself, and this would be the new way of warfare.

The reason that the American airships were such a threat was that the Americans had a monopoly on the world supply of helium. Although this is a safe way of lifting dirigibles, helium is less powerful than

AMERICAN STYLE DIRIGIBLE

hydrogen peroxide because it gives less lift and is not as fast. Hydrogen peroxide is much more flammable and therefore very dangerous. Having no other choice, the Japanese were willing to take the risk, and because of their willingness for self-sacrifice, they were able to get higher lift, just as the Germans had done in World War I, but they were able to develop their craft far beyond what the Germans did.

Although America's dirigible efforts were discontinued due to various mishaps and accidents, this is a cover story for what actually happened. The Japanese were able to sabotage all of the American airborne carrier vessels so that they all went down, Hindenburg style. They were able to do this and take dominance over airship technology in the skies as a result of what had happened during America's invasion of the Philippines in 1899.

In 1899, the Americans had invaded the 7,100 islands of the Philippines, launching an invasion which killed no less that three million Filipino Muslims.* During that time, many Asians developed an intense hatred for Americans, and this is why, to this day, Muslims share a similar hatred. As Japan had a direct maritime border with the Philippines at that time, the American invasion of those islands would have been the equivalent of the Japanese invading Mexico or Canada and slaughtering millions of Mexicans or Canadians.

After the invasion, the American Navy became saturated with Filipino stewards. I learned this from my father who had served in the U.S. Navy. You can research this, and it can be verified. They had taken all these Filipinos and literally enslaved them. He told me that there are Filipinos who died here (in the United States) in the 1980s and 1990s who had been slaves to the United States, having been shipped to America to serve the U.S. Navy and were never granted citizenship. These Filipinos were easily infiltrated by Japanese espionage agents that today's Americans might very luridly refer to as "ninja", and they sabotaged the entire naval air fleet program and sent it flaming out of the skies. In other words, Japanese infiltrators amongst the Filipino stewards and slaves saw to it that the airships were blown up. After that, the American naval air ship program died. After the third such

---

*Following the Philippine invasion during the Spanish American War, a national American newspaper, *The Argonaut*, was talking about how wealthy the Filipino Island chain was and stated that the only problem was that it was infested with Filipinos. They said all we need to do is kill all of them and American capital can turn the Philippines into a paradise. Another indicative example of such genocidal racism is a report from the U.S. Army from Manila of February 5, 1899 which reads: "In the path of the U.S. Army regiment and Battery D of the 6th artillery, there were 1,008 dead niggers and a great many more wounded. We burned all of their houses and killed all the women and children."

THE CHINESE WAR

American dirigible disaster, the Japanese picked up where the Americans had left off and were able to perfect the technology of the dirigible which was basically a very large hydrogen balloon surrounded by eight subsidiary balloons, all covered with a single silk "skin-sheet" for the sake of aerodynamics. Together, these nine balloons measured well over half a thousand feet in diameter. They were so huge they were called super-dirigibles, and they featured a remarkable lifting capacity of over 250 tons for each balloon. As an aggregate, they lifted quite a bit, including several tankettes at a time that they could drop into Chinese walled cities. The dirigibles you see here are computerized renditions of true-to-life illustrations I did based upon on the photographs I was assigned to destroy.

ABOVE IS A RENDERING OF THE TYPE OF CRAFT THAT CRASHED OVER ROSWELL. THE STRUCTURE AT THE BOTTOM COULD HOLD VARIOUS AIRCRAFT. THE MOORING LINES COULD ALSO HOOK PARASITE PLANES WHICH WOULD BE PULLED UP INTO THE INTERIOR OF THE OVERALL STRUCTURE.

THE ROSWELL DECEPTION

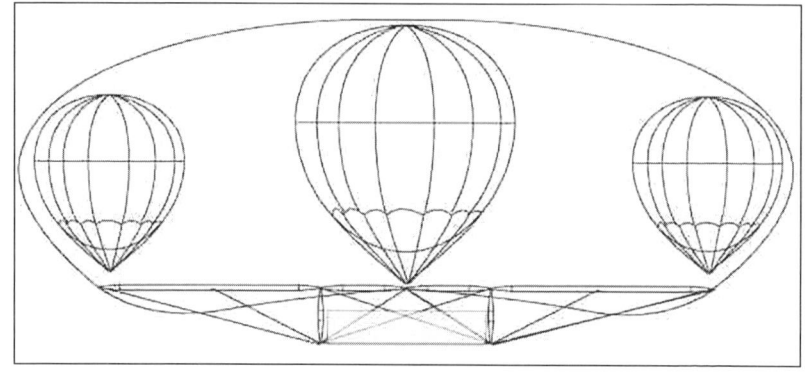

THE ILLUSTRATION ABOVE SHOWS SPACE WITHIN THE CRAFT WHICH WAS USED TO STORE PLANES, TANKS, AND OTHER WEAPONS OF WAR. THERE WERE ACTUALLY MORE BALLOONS INSIDE, BUT THIS ILLUSTRATION IS FOR DEMONSTRATION PURPOSES ONLY.

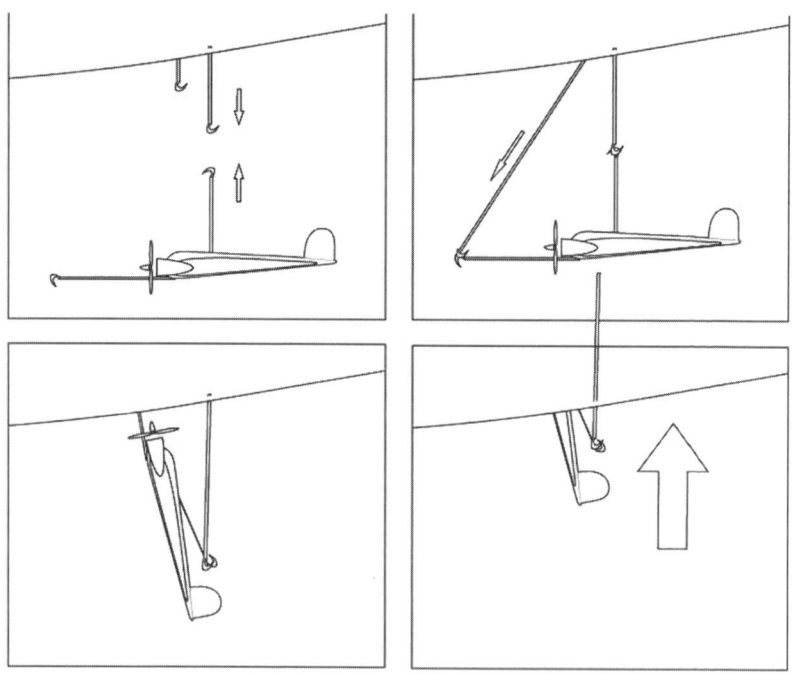

THE DIAGRAM ABOVE DEMONSTRATES HOW PARASITE PLANES COULD BE HOOKED AND PULLED INTO THE INTERIOR OF THE DIRIGIBLE.

THE CHINESE WAR

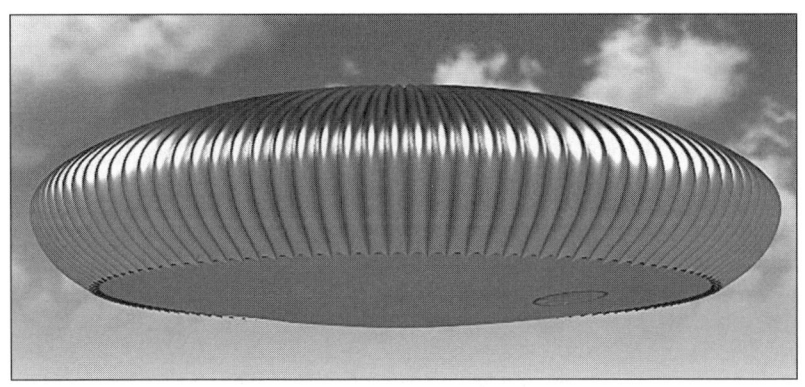

Above shows how the dirigible could be covered from the bottom. The folds could be manipulated to open up the containment spaces that were needed to store parasite craft, tankettes or other weapons of war.

Above is a closer look at the underside of a "hamburger" dirigible which shows a "flying pancake" (plane) which was a Japanese modification of an American Vought V-173.

ABOVE IS AN ITALIAN MANTA BOMBER ATTACHED TO THE DIRIGIBLE. SEVERAL OF THESE WERE MADE IN ITALY AND SHIPPED TO JAPAN.

The Japanese built extremely small tanks that are known as tankettes. These would be considered worthless on the modern European battle theater but they were perfect for China as the Chinese had no armor by which to counter them. To overcome the enormous walls and fortifications of the Chinese warlords, the Japanese would airdrop tankettes into the middle of the city square along with soldiers who would use the tankettes as a vanguard. These tanks were too small for a normal Caucasian American to fit into. They were only suitable for those Japanese who might average about four feet in height.

The Japanese continued to perfect their dirigibles, mounting banks of M99 light machine-guns with which the Japanese killed more Allied soldiers than with any other weapon in the Second World War. The dirigibles were also mounted with the Japanese version of the German 88 heavy anti-aircraft/anti-tank artillery gun. The purpose was to get these armaments in the air where they could use them to maximum effect and pound Chinese armies into rubble.

Borrowing from the American concept, the Japanese mounted their parasite aircraft through a circular rim inside the dirigible via a hook system. In launching, the pilot would enter the aircraft in such

THE CHINESE WAR

JAPANESE TANKETTE

a position as to fly prone on his stomach. As the Japanese were so short, they could accommodate that position for hours, much longer than a Caucasian could comfortably do so, aside from the fact that the latter could not even fit inside of such a small aircraft. In retrieval, the aircraft would be hooked and then dragged into the mother ship vertically, like Peking ducks on a rack system, where the pilot could then exit. That is how they actually hung the planes on the inside of the outer shell. The outer shell of the balloon-aggregate consisted of rubberized silk so that the space in between the separate balloons within the shell, where the personnel conducted launch and retrieval operations, would be sheltered from high winds in addition to it facilitating smooth sailing. Rubberized silk has a retractable quality whereby it returns back into its original shape when it is bent or distorted. At Roswell, this was later misidentified as retractable metal.

As the Japanese dominated the silk industry, the Americans did not even have the materials to build such ships. So, with all the silk the Japanese had, they were able to create unconventional bio mimetic craft that could float in the sky. Bio mimetic means that they were mimicking the natural form of various animals, and this is why people said the shapes of these unconventional craft appeared biological to the eye. As they were very alien looking, this would encourage many Americans, in the end, to misinterpret such craft as extraterrestrial phenomena.

These super-dreadnoughts were so enormous that huge containers were built that were capable of carrying troops, paratroopers, ammunition, food, or anything else needed. You can see how tanks or bombs could be dropped onto Chinese cities or villages. This is how the Japanese maintained control of the Asian mainland.

There was even one special dirigible that was created for Emperor Hirohito himself. He had his own compartment that even included a bed. Everything was light weight and made of bamboo so that it would not take up too much weight. This was the super battle dirigible "Mikado", and it also featured a plasma cannon. The static electricity in the atmosphere is so incredible that you can basically generate thousands and thousands of volts and the Japanese were trying to harness this to turn it into a weapon because some of the dirigibles they were creating were so enormous. The plasma cannon would absorb all of the static electricity that was collected around the skin of that dirigible and would discharge it in artificial lightning bolts. This weapon was iffy at best and its behavior was highly erratic, and it was removed when the Battle of Los Angeles took place, an event that will be addressed in the next chapter. They removed the plasma cannon because they had determined that the appearance of the craft alone would distress the Americans enough as it was approximately 300 yards long in size. The end result of that adventure to Los Angeles, however, resulted in the Americans killing thousands and thousands of their own men.

These dirigibles could absorb a lot of damage. If one, two or even more balloons were taken out, there was still enough lift to keep the crew and craft afloat. Even if the main balloon was taken out, a few of the subsidiary balloons could still sustain flight and return the crew back to Japan on the Kuroshio Current. Injured balloons were dumped, and the U.S. Navy sometimes picked up huge rubberized silk balloons that they could not identify until later in the war. The Chinese, who were more than informed about these super-dreadnoughts, told the Americans what they were dealing with, and this turned out to be an important piece of intelligence that greatly influenced World War II.

While the Chinese war had causes of its own, it served as a perfect testing ground for what would be unleashed a few months after the Pearl Harbor incident.

THE CHINESE WAR

ABOVE IS THE MIKADO, THE EMPEROR'S SUPER-DIRIGIBLE, WHICH INCLUDED SLEEPING QUARTERS AND A PLASMA CANNON. THE (FRONTAL) BOW DISPLAY OF THE IMPERIAL "KIKUMON (CHRYSANTHEMUM SEAL)" IS RENDERED IN CORK.

ABOVE IS THE MIKADO WITH THE PLASMA CANON IN ACTION.

Above is the Italian Manta Bomber which could be attached to the dirigible as a parasite craft. At the end of the tail is a hook to attach it with.

CHAPTER ELEVEN

The United Nations

Although it is broadly recognized as an international organization, the United Nations and its charter are not well understood with regard to the ramifications of World War II and the position of the United States and Japan in particular.

The initial formation of the United Nations was on January 1, 1942, exactly three weeks and three days after the surgical strike on Pearl Harbor. When it was formed, a charter was established in lieu of a constitution wherein it was formed as a military organization of war as per Title 42 of the United Nations Charter. Title 42 specifically states the following.

> "Should the Security Council consider that measures provided for in Article 41[*] would be inadequate or have proved to be inadequate, it may take such action by air, sea, or land forces as may be necessary to maintain or restore international peace and security."

The initial formation of the United Nations was at the Presidio military base in San Francisco which it used as its Pentagon. From the Presidio, the duty of the United Nations was to coordinate all of the Axis resistance armies that existed behind the front lines of Axis occupation in order to fight the Axis powers. The resistance armies of Denmark, France, Holland, the Philippines, and all other occupied or unoccupied of the half a hundred Allied Nations throughout the world were all coordinated through the Presidio Western Defense Command Center in San Francisco under the specific direction of the United Nations.

Although the peace branch of the United Nations was opened in 1945 in New York, it was the war branch at the Presidio where the peace treaty between the United States and Japan was ultimately signed on September 8, 1951. This is where I worked as a Department of Defense research librarian for nearly a decade, having already read everything I could in the previous decade in the Presidio library, much of which was open reading for military dependents such as myself.

---

[*]Article 41 of the United Nations Charter is as follows: "The Security Council may decide what measures not involving the use of armed force are to be employed to give effect to its decisions, and it may call upon the Members of the United Nations to apply such measures."

In terms of everything I was exposed to, in terms of classified information as a genuine military librarian, I can tell you that, as I have previously explained, we were at war well before Pearl Harbor took place, both in Asia and also unofficially in the Atlantic. When the war came to an end, it was actually at a far different time than most people realize, and an overwhelming number of relevant documents pertaining to such were ultimately destroyed. I destroyed quite a few of them as a Department of Defense librarian under orders. I incinerated them so they would never go to the media nor to academe. Consequently, everything the professors and authors of history write is based upon very incomplete and quite faulty information. Nevertheless, a lot is still out there. You can find that the Japanese-American peace treaty, referred to above as the *Treaty of San Francisco*, was not signed until 1951 and did not go into effect until Emperor Hirohito's birthday on April 28, 1952. Up to that point, we were still legally at war, and a war which very much ended in Japan's favor. I will go into the circumstances of this in more detail at a later point in this book.

The attack on Pearl Harbor definitely changed the landscape of a world war that was already in progress, despite the fact that the Congress had not officially declared it. Less than two months after the inception of the United Nations, President Roosevelt ordered the internment of anyone whose ethnicity could be construed as being that of an enemy alien. This included the confiscation of assets of actual citizens of the United States who were of Japanese descent, even those who hailed from families who had been American citizens for three generations, and placing them in internment camps, a "polite" word for a concentration camp.

Only eleven weeks after the Pearl Harbor attack, and less than a week after President Roosevelt's order to intern all ethnic Japanese, Emperor Hirohito unleashed his dirigibles in what has become known as the Battle of Los Angeles

CHAPTER TWELVE

THE BATTLE OF LOS ANGELES

On the evening of February 23rd, 1942, Franklin Delano Roosevelt was giving one of his famous fire side chats where he would attempt to brief the public and reassure them. During the Great Depression, these were generally very comforting to an impoverished and very disturbed population. It gave them the impression that someone actually cared about their plight.

While Roosevelt gave his chat on that night, a Japanese submarine had surfaced off the California coast near Santa Barbara, just about 90 miles north of Los Angeles. According to news reports, it fired between 12 and 25 shells of ammunition upon the Ellwood Oil Field near the small town of Goleta. With no human casualties, the damage was limited to an oil derrick, a pumping station and a pier.

With both countries in a full-blown declared war at that point, this was not a serious attack. The Japanese were testing the American response before they would unleash their fleet of aircraft in what would become known as the Battle of Los Angeles. There was no response by the Americans other than a limited coastal blackout which did not even extend as far as Los Angeles.

As this was a clear signal that the Americans were not prepared, the Japanese began what was a planned psychological operation on the United States that was specifically directed over the city of Los Angeles. Just two months and seventeen days after the preemptive strike on Pearl

115

Harbor, Japan's covert black-budget or "Black Sun" operation, under the emperorship of Hirohito, had ordered the super-dreadnought dirigibles to float across the Pacific Ocean using the Pacific jet stream and float down the coast, eventually drifting over the city of Los Angeles.

    The objective of the Japanese was to display force and put the fear of God into the Americans. In this respect, Hirohito was a combination of Napoleon Bonaparte and Hannibal Lecter because he was very well aware that the Americans knew he had stolen so many biological weapons. Despite this, he did not load any biological weapons because they were in liquid form and therefore extremely heavy. The Battle of Los Angeles was strictly a psychological maneuver. The dirigibles simply rendered themselves visible over Los Angeles, refraining from employing their smoke distribution cover. Their mission was to float

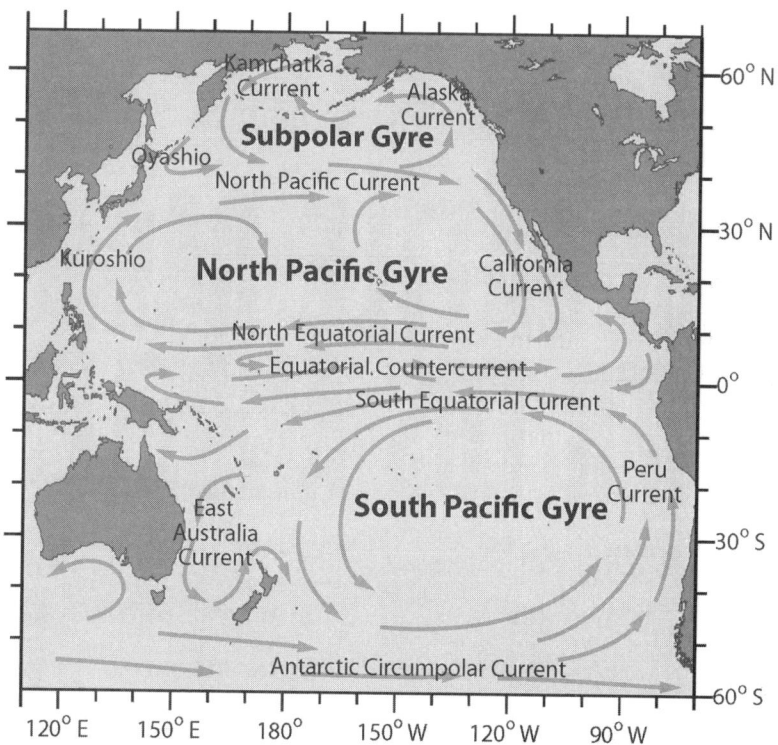

ABOVE IS A MAP OF THE KUROSHIO OR JAPANESE CURRENT THAT RUNS UP THE JAPANESE COAST AND INTO OTHER CURRENTS THAT FACILITATE PASSAGE TO CALIFORNIA AND A RETURN TO JAPAN.

down the California air current and then use the Equatorial air current to float back to Japan which, in those days, stretched all the way down to the Equator because the Japanese retained all of Micronesia and occupied New Guinea.

The U.S. Army responded to the Japanese with air raid alarms to warn the public of violation invasion of American air space and attempted to counteract the invasion. Anti-aircraft guns fired over 14,000 rounds at their flying targets, but none of the rounds could reach the dirigibles because their flight ceiling capability was so high. As a result, round after round of the ammunition fell to ground and took the lives of many civilians thereon who had come outside to witness the invasion. The dirigibles also deployed Japanese parasite fighters to intercept anti-aircraft fire, thus further enabling the motherships to evade the majority of the rounds. The invasion and air raid lasted several hours in total, extending into the 25th of February. This resulted in some Japanese

LOS ANGELES EXAMINER HEADLINE REPORTING
PLANE DOWNED ON VERMONT AVENUE

LONG BEACH PRESS TELEGRAM OF FEBRUARY 25, 1945

THE BATTLE OF LOS ANGELES

planes being shot down with newspapers giving street addresses of where the planes actually crashed.

The *Long Beach-Press Telegram* of February 25, 1942 ran the headline, "GUNS BOMBARD SKYS AS STRANGE PLANES BLACKOUT SOUTHLAND IN FIRST 'RAID'". The headline for the *L.A. Times* read, "ARMY SAYS ALARM REAL". The *Los Angeles Examiner* headline read, "AIR BATTLE RAGES OVER LOS ANGELES" with a sub-headline which read, "One Plane Reported Down on Vermont Avenue by Gunfire". Specific addresses were also given to where other planes had been shot down.

When I was in high school during the 1980s, radio station KGO in San Francisco transmitted a radio broadcast about the Battle of Los Angeles. They received two hundred letters and dozens of phone calls in response. Retired air raid wardens called in and indeed verified that those were Japanese planes. Dan Tryon of Redding, California was ten years old at the time and a magazine peddler. He said that he picked up pieces of bloody Japanese parachute harnesses and pieces of Japanese planes. When he went back the next day, everything was all gone and the soldier guarding the field said it was "aliens". The fix was on instantly.

The Office of War Information orchestrated an immediate and extensive cover-up which is still very much in force to this day. If you look at the photographs of the Battle of Los Angeles (see next page), you will see an image alteration technique used back in the 1960s. Inside those searchlight beams of the Battle of Los Angeles, an old fashioned manual version of "cut & paste", which was the "Photoshop" of its day, was used to deliberately insert a "saucer-shape" into the image. If you watch the footage of the battle of Los Angeles, however, you will not see any object in the light beams, and if you listen to the radio transmission, you will hear it said that it was a Japanese invasion with Japanese planes. And as we just said, the specific locations of the downed Japanese planes were even stated in the newspapers.

This propaganda is still promoted heavily through Hollywood movies such as *The Battle of Los Angeles*, produced in 2011 during the height of the Iraqi conflict. The United States military entertainment complex is so dedicated to reinforcing this disinformation about what happened that they brought back Apache and transport helicopters from the Middle East to shoot the footage for this movie, using real marines to pilot those helicopters, and the military also paid for those helicopters to be shot in that film. In other words, the production company never paid for it and was being supported and funded to help promote the agenda that aliens were the cause. In another military

ABOVE IS THE DOCTORED PHOTO OF A CONFLUENCE OF SEARCH-LIGHT BEAMS DURING THE BATTLE OF LOS ANGELES. THE PHOTO BELOW IS OF THE ACTUAL "SHEFFIELD UFO" THAT WAS CUT AND PASTED INTO THE SEARCHLIGHT BEAMS ABOVE IN ORDER TO MAKE YOU THINK THE BATTLE OF LOS ANGELES WAS A UFO INVASION.

THE BATTLE OF LOS ANGELES

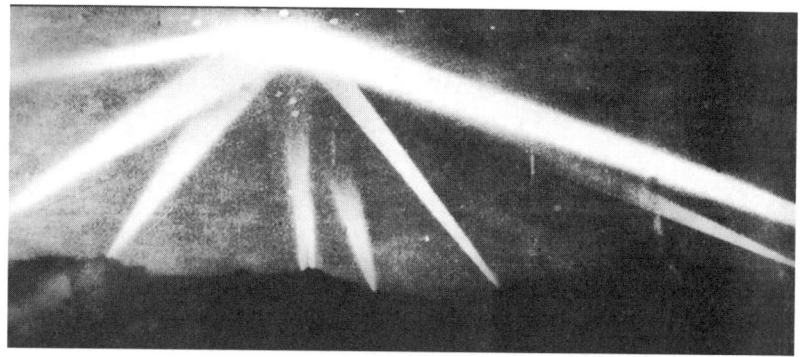

ABOVE IS AN ORIGINAL PHOTO FROM THE BATTLE OF LOS ANGELES

ABOVE IS ANOTHER UNDOCTORED PHOTO IDENTIFIED AS "BEAMS FROM COAST ARTILLER SEARCHLIGHTS SPLIT THE NIGHT SEEKING AIRCRAFT OF LOS ANGELES NIGHT OF FEBRUARY 24-25, 1942".

THE ROSWELL DECEPTION

BELOW IS A PHONY TOP SECRET DISPATCH FROM THE PRESIDENT THAT WAS "LEAKED" SO AS TO FURTHER THE PSYCHOLOGICAL CAMPAIGN AGAINST THE PUBLIC.

TOP SECRET

Top Secret

gm 25
February 27, 1942

THE WHITE HOUSE
WASHINGTON

February 27, 1942

MEMORANDUM FOR

CHIEF OF STAFF OF THE ARMY

I have considered the disposition of the material in possession of the Army that may be of great significance toward the development of a super weapon of war. I disagree with the argument that such information should be shared with our ally the Soviet Union. Consultation with Dr. Bush and other scientists on the issue of finding practical uses for the atomic secrets learned from study of celestial devices precludes any further discussion and I therefor authorize Dr. Bush to proceed with the project without further delay. This information is vital to the nation's superiority and must remain within the confines of state secrets. Any further discussion on the matter will be restricted to General Donovan, Dr. Bush, the Secretary of War and yourself. The challenge our nation faces is daunting and perilous in this undertaking and I have committed the resources of the government towards that end. You have my assurance that when circumstances are favorable and we are victorious, the Army will have the fruits of research in exploring further applications of this new wonder.

You may speak to me about this if the above is not wholly clear.

F. D. R.

propaganda flick, *Independence Day*, released in 1996, you will see American and Japanese troops mobilized to fight against an evil outer space invasion, a theme that is even more intensely re-empahsized in the movie *Battleship* in 2012. This is how the United States covers up their own mistakes and crimes and the extent to which they will do so. Further, if you go to youTube, you will find hundreds of channels and people still promoting the alien agenda and still believing to this day that it was aliens from outer space who invaded Los Angeles, having no concept or notion it was the Japanese who invaded during the Second World War. As it occurred during wartime, it should make logical and common sense that it was the Japanese, but this is how effectively propaganda and disinformation, pervasively promoted, works to dupe people in order to hide the true reality. A prime example is a phony top secret dispatch from the President that was "leaked" so as to further the psychological campaign against the public, and even against their own federal subordinates, that the attack was from extraterrestrials.

You can better understand the Government's motive for hiding this information when you learn the actual response to the presence of Japanese dirigibles and airplanes buzzing Hollywood.

The Americans were well aware that the Emperor was developing biological weapons because of their deployment against the Chinese on the mainland. So, after the Battle of Los Angeles was over, the American high command was asking the question, "If the Japanese were not dropping bombs on us, what were they doing?"

Knowing well the history of Hideyo Noguchi, they readily concluded, quite incorrectly, that the Japanese must have put the yellow fever virus into an aerosol form and sprayed the contagion over Los Angeles. The immediate result of this errant conclusion was that the military locked down Los Angeles into a quarantine zone and did not allow anyone to leave the city. Further still, in a misguided and crazy attempt to protect their own military men and soldiers from dying, they injected their own troops with a yellow fever vaccine that was not approved by the Food and Drug Administration (FDA). In other words, it did not work. It also had disastrous side effects.

After developing a vaccine out of aborted babies and adult cadavers, the application of it unleashed a horrendous outbreak of hepatitis amongst the military, killing over 50,000 of their own men with over 300,000 additional personnel being severed from military service as a result of permanently debilitating illness. You can learn about this in accessible history.

The incompetence of the United States military high command in issuing that order to vaccinate the troops resulted in an extensive cover-up for the express purpose of hiding their own mistakes that resulted in massive deaths and mass-mustering out of the service of their own men. This is why there has been tremendous censorship concerning the Battle of Los Angeles and why it has been covered up by promoting it as an "alien" phenomenon. The Government does not want the public nor their own troops to know that they have historically treated them as expendable fodder. The term "alien" actually refers to Asian Pacific Islanders who were considered "alien enemies" of the United States. Just before the Battle of Los Angeles, Franklin Roosevelt issued Executive Order 9066 which established internment camps for enemy aliens. The nuance with the word "alien" has been used ever since, and that is why the battle of Los Angeles has been promoted for over 70 years as an "alien" invasion from outer space.

The Battle of Los Angeles was the greatest victory of World War Two by the Japanese because, without dropping a single bomb and without firing a single round, Hirohito had not only struck abject fear into the hearts of his enemy, they panicked to the point of rampant self-destruction, killing and/or permanently disabling their own soldiers by the hundreds of thousands. The calculation and precision of the Emperor was without reproach, at least in terms of its effectiveness. It would put an imprint on the American high command that would eventually enable Hirohito to secure his own longevity and a peace that would enable his country to prosper like it never had before.

When the *Treaty of San Francisco* was officially signed in 1951, thus ending the war between the Allies and Japan, Emperor Hirohito handed the yellow fever vaccine over to the Americans and said words to the effect of: "Here — you can have this since you killed so many of your own men with something with which you didn't know you were working."

The following pages include artistic renditions of what the Emperor's fleet looked like during the Battle of Los Angeles, including the parasite planes that could be detached and fly independently. Once again, the fleet did not contain any biological weapons during their invasion of L.A. air space, but the Army sure thought that they did.

THE BATTLE OF LOS ANGELES

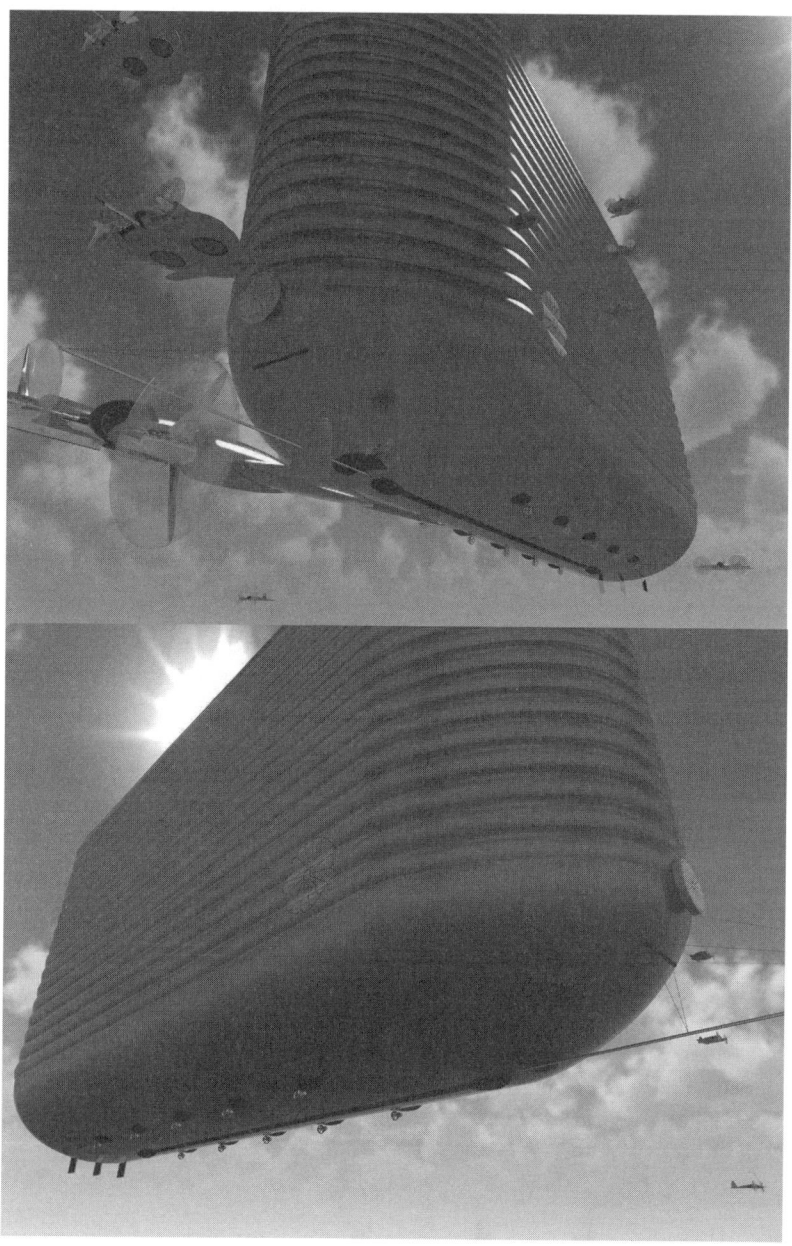

UNDERSIDE VIEWS OF THE EMPEROR'S MIKADO
(NOTE THE CORK CHRYSANTHEMUM SEAL ON THE FRONT)

THE ROSWELL DECEPTION

The Mikado and three "Flying Hamburger" super-dirigibles

modified Vought V-173's, also known as "Flying Pancakes"

CHAPTER THIRTEEN

THE OFFICE OF WAR INFORMATION

The reason you have not heard of the data I have presented to you in this book is the Office of War Information (OWI). It is the greatest enemy of your mental freedom or your capacity to think so as to make correct decisions as an informed electorate. Although the OWI was disbanded after the cessation of World War II hostilities, what I have said here also applies to its predecessors and successors.

While it is understandable that any political entity will act to prevent leaks from its own private counsel and strategic planning, the Office of War Information has acted in such a manner so as to have induced the "GI generation" into an unconditional hatred of genocidal mania that has been inoculated into the American electorate so as to create a reverence for an executive dictatorship that is nationally self-destructive. It is the OWI that is behind the direction, limitation and censure of all major mass media outlets: radio, television and film, thus creating a fourth estate of media suppression cordoned in by the military minefields of DARPANET, which you know as the internet.

Within two weeks of the Pearl Harbor attack, on December 19, 1941, President Franklin D. Roosevelt signed Executive Order 8985 which established the Office of Censorship in order to censor all communication coming into and going out of the United States. Within six months, Roosevelt issued Executive Order 9182 which established the Office of War Information, consolidating the Office of Facts and Figures, the Office of Government Reports, and the Division of Information of the Office for Emergency Management. Both the Office of Censorship and the Office of War Information answered directly to the President himself and were concerned with suppressing information, the latter utilizing propaganda to control how the war should be viewed, both on the international front as well as domestically. President Roosevelt personally appointed Elmer H. Davis to head up the Office of War Information, and what follows is a famous quote from Davis with regard to influencing people's minds.

> "The easiest way to inject a propaganda idea into most people's minds is to let it go through the medium of an entertainment picture when they do not realize that they are being propagandized."

Elmer Davis was an editor for *Adventure*, an early pulp magazine which spawned the Adventurers' Club of New York, a circle of influential people who were either key military and government figures or closely tied to them. Davis was already a long time trusted team member of such circles prior to this appointment. At that time, the pulps were already being used to steer the population with regard to technology and new inventions.

The Office of War Information was formed under the influence of a genocidal mentality that had a long history in America, dating back to smallpox blankets, Jack London's *The Unparalleled Invasion* and the American Army-Navy Flu unleashed by Woodrow Wilson. At the time of the OWI's inception, there were currently two effective propaganda models upon which it patterned itself from, both with a genocidal agenda.

The first of these models was the state-sponsored work of H.G. Wells, a name that most will recognize as the author of *The Time Machine*. A propaganda officer for the British government, Wells was far more than just a best-selling author and had an enormous impact on the world around him, one of his primary themes being the purging of the human race, effectively bleeding out the less fit so that only the quality people would be left alive after the majority of the population had been exterminated. A leading member of the influential and mysterious Fabian Society, Wells was an ideological communist, socialist, and Stalinist who also served as a goodwill ambassador on behalf of England. One of his tasks in this regard was to visit Russia and interview Stalin. His article referred to the homicidal dictator as "fair and candid".

The extreme views espoused by Wells centered around an anglocentric agenda and mindset for an apocalyptic sense of total war that it was thought necessary to manifest evolutionary progress, the reasoning being that this concept of genocidal warfare would somehow gel Darwinian advancement for the human species. He wrote books on how this could be done on Earth, between nations and opposing civilizations.

Published on the eve of the Twentieth Century in 1898 – the year Adolph Hitler himself was born – H. G. Wells' *The War of the Worlds* is much more than just a seminal work of science fiction. It is also a kind of Darwinian morality tale, and at the same time, a work of singular prescience. In the century after the publication of his book, scenes like the ones Wells imagined became a reality in cities all over the surface of the Earth as modern mass industrialized conflict and the transitional ascendancy of Wells' own promulgated Totalitarianism out of Eurasia was expedited by the degenerate West via their Alliance against Axis Civilization. In this respect, the United States Government fielded a football team that was all offense and no defense, doing so by concentrating all of their intellectual and material assets on warring against the most vulnerable elements of foreign populations (i.e. women, children, infants, and the elderly). By conducting war in this manner, American leadership echoed Wells' Martians by leaving their own continental homeland wide open for threat-projection of a fatal bacterial blow by the Japanese Monarchy and its war machine.

The American equivalent of H.G. Wells was E.E. "Doc" Smith, a chemist and celebrated science fiction author who wrote the *Lensman Series*. Professionally, Smith made mixes for doughnuts and cereals, but as a chemist, he is best known for his paper, "The effect of bleaching with oxides of nitrogen upon the baking quality and commercial value of wheat flour". In other words, his work is a substantial part of the legacy of nutritionless and even toxic white flour.

The *Lensman Series* of books, which he finally completed while in military service during World War II, were far more influential than anyone might realize. The *Lensman Series* was a story about exterminating different races of people on the basis that they were inherently evil by reason of their genetics. Smith presented this on a galactic scale where even entire planets should be wiped out and entire races should be exterminated without remorse.

To give you an idea of how influential he was, Smith was a guest of honor at the Second World Science Fiction Convention and was inducted into the first Fandom Hall of Fame at the 21st Worldcon. Long after World War II had reached a cease-fire, Smith received a letter via John W. Campbell, relayed from Admiral Chester Nimitz himself, who acknowledged that he had emulated E. E. Smith's "Display Tank", as outlined in his science fiction stories, which was called "Executrix", a massive computer holographic display tank which could portray entire galactic engagements. Admiral Nimitz admitted that the U.S. Navy had based their entire concept of communications between aircraft carriers upon the visionary ideas presented by E. E. "Doc" Smith. If he was influencing the operational war machine to that level, you can imagine how he was influencing the ideology of entire generations who were raised reading his books.

The second contemporary model that the OWI was based upon, however, was not an American model at all. It was that of Joseph Goebbels' Reich Ministry of Public Enlightenment and Propaganda.

Goebbels was recognized as one of the only German generals who never lost the war. The reason for this reputation is that it was his stated job description to totally mobilize the German citizenry into an effort of total war and self-sacrifice until the very bitter end. In this effort, he succeeded quite admirably; and in this respect, the effectiveness of his propaganda ministry cannot be denied. It was essentially there to instill, mobilize and inspire. The United States, however, had a very different approach.

While Goebbels set out to vilify the Jews, using ridiculously unflattering characterizations in cartoons and other media, he also built up the self-esteem of the German population by the repeated declaration of Aryan superiority. This, however, would not work for the Americans. While Roosevelt clearly used the term "Aryan superiority" and believed he himself was part of an elite race, he had no such conviction for the majority of the American population. They were a melting pot and a rabble in his eyes. Instead, the Office of War Information concentrated on dehumanizing and vilifying their enemies, particularly the Japanese.

Prior to Pearl Harbor, most Americans were isolationist and wanted nothing to do with foreign wars. It was only through the manipulation of events which led to the Japanese attack that Roosevelt was able to muster any popular support for a war effort. Even then, most people did not understand what the conflict was really all about. This is where the Office of War Information came into play. The solution was to dehumanize the enemy and turn them into monsters. In other words, you were not killing actual people but rather vermin.

The mentation pattern that had been induced into the GI generation was taken full advantage of by the OWI. Admiral William Halsey, who commanded the U.S. Naval forces in the South Pacific throughout World War II, articulated that the official mission of the United States military was to "Kill Japs, kill Japs, kill more Japs," and he vowed that, "Before we're through with them, the Japanese language will be a language spoken only in hell."*

---

*With his salty language and aggressive manner, Admiral Bull Halsey was the Navy's equivalent of fighting General George Patton. Another example of Halsey's audaciously rank behavior, which eventually led to him looking extremely foolish, concerned his bragging vow that he would one day ride Emperor Hirohito's white horse through the streets of Tokyo.

The Emperor's identification with his white horse was a carefully cultivated image, the history of which went back to the 19th Century when President Ulysses Grant had gifted the Meiji Emperor with a beautiful jet black stallion. This equestrian alliance continued into the following century when another beautiful stallion, Shirayuki (meaning "White Snow"), was shipped to the Emperor from California. Hirohito was often shown in photos and film clips riding his white stallion in front of his troops.

As the Emperor on his horse became such an iconic symbol, Halsey's vow became a rallying cry in America and was even used in a campaign to sell war bonds. The theme was that the United States was going to win the war and take Hirohito off his high horse.

At what most believed was the end of the war, there was a public demand to see Halsey ride the Emperor's white horse through the streets of Tokyo. While the Admiral did ride a horse, it was not the Emperor's. Instead, another *(continued on next page)*

EVEN THE AMERICAN PRESS WAS HAVING FUN AT ADMIRAL HALSEY'S EXPENSE IN THE ABOVE DRAWING WHICH WAS FEATURED ON A MAGAZINE (UNKNOWN) COVER.

(continued from previous page) white horse was supplied by Major General William Chase, the commander of the First Calvary Regiment. After a graceless and very short ride on the cavalry camp outside of Tokyo, the relieved Halsey dismounted and said "Please don't let me alone with this animal. I was never so scared in my life."

The Japanese, however, used Halsey's ride as their own internal propaganda coup. The clumsy ride was filmed and broadcast across the country, much to the amusement of the Japanese public. Further, the American press had no choice but to tell the clamoring American public that the Admiral "will never ride that white horse (the Emperor's) except by imperial invitation."

This mindset extended down to schools where the following example was used to explain the war to young students and what Japanese were like. It was included in *Leatherneck* magazine, published by the US Marine Corps in 1945. The words in the article are repeated below.

Louseous Japanicas

The first serious outbreak of this lice epidemic was officially noted on December 7, 1941, at Honolulu, T. H. To the Marine Corps, especially trained in combating this type of pestilence, was assigned the gigantic task of extermination. Extensive experiments on Guadalcanal, Tarawa, and Saipan have shown that this louse inhabits coral atolls in the South Pacific, particularly pill boxes, palm trees, caves, swamps and jungles.

Flame throwers, mortars, grenades and bayonets have proven to be an effective remedy. But before a complete cure may be effected the origin of the plague, the breeding grounds around the Tokyo area, must be completely annihilated.

Louseous Japanicus

"The first serious outbreak of this lice epidemic was noted on December 7, 1941 at Honolulu, T.H. To the Marine Corps, especially trained at combating this type of pestilence, was assigned the gigantic task of extermination. Extensive experiments at Guadalcanal, Tarawa, and Saipan have shown that this louse inhabits coral atolls in the South Pacific, particularly pill boxes, palm trees, caves, swamps and jungles. Flame throwers, mortars, grenades, bayonets have proven an effective remedy. But before a complete cure may be effected by the origin of the plague, the breeding grounds in the Tokyo area must be completely annihilated."

In 1943, well before heavy fighting had begun in the Pacific Theater, a U.S. Army poll showed that over half of every GI interviewed thought it would be necessary to kill every single Japanese person on Earth and that this war was a war of absolute extermination.

THE OFFICE OF WAR INFORMATION

14. The Western perception of the Japanese as "little men" or "lesser men" meshed easily with images of the enemy as primitive, childish, moronic, or emotionally disturbed. This graphic, originally published in the *Detroit News* on the occasion of Japan's surrender in August 1945, reached a much larger audience when it was reprinted in the Sunday *New York Times*.

ABOVE IS ANOTHER CREEPY EXAMPLE OF THE DEHUMANIZATION OF JAPANESE BY THE AMERICANS.

Another example of the mindset in vogue can be seen (on page 134) in *LIFE* magazine's picture of the week of May 22, 1944 featuring a full page with Natalie Richardson of Phoenix, Arizona being photographed with a Japanese skull she was using as a paper weight. It had been sent to her by her boyfriend, Lt. Wemyss, and it was autographed by his entire squad.

This full page spread of Natalie Richardson basically became the vanguard for many other atrocities involving Japanese body parts. Every single jeep in the Pacific bore samples that were similar to the photo on page 135. The skulls pictured there are of a husband and a wife that were mounted on the hood of a jeep. Every jeep in the Pacific pretty much had hood ornaments like that. The Japanese had colonized the Pacific,

THE ROSWELL DECEPTION

THE CAPTION FROM *LIFE* MAGAZINE READS: "ARIZONA WAR WORKER WRITES HER NAVY BOYFRIEND A THANK-YOU NOTE FOR THE JAP SKULL HE SENT HER."

and they had thousands of settlers. The U.S. Army exterminated them and used their body parts and would paint them. A Japanese soldier's head would be mounted on a tank hood or serve as a hood ornament, and they would also send these home. Necklaces were made of Japanese teeth, and letter openers were made of Japanese jaw bones. Even President Roosevelt received one and kept it on his desk. These are all blatant examples of the dehumanization that occurred on behalf of the Americans.

    A year after the photograph with Natalie Richardson, *LIFE* magazine ran an article on June 11, 1945 which stated, "Total war calls for total defeat. That means that the war making power of every Japanese resource, every Japanese man, woman and child must be destroyed." Keep in mind that *LIFE* was heavily vetted by the OWI and was merely serving its purpose in articulating official military objectives.

THE OFFICE OF WAR INFORMATION

SKULLS OF A JAPANESE COUPLE MOUNTED ON AN AMERICAN JEEP

This rhetoric and preoccupation with extermination went so far that U.C. Berkeley began to preserve specimens of Japanese body parts in the anthropology department because they were convinced that the entire Japanese race would be exterminated. They were asking soldiers to bring in complete bodies of women and children back to the States so they could be preserved, pickled and basically stuffed for showing in museums because they figured future generations would never see another Japanese.

To this very day, one of the biggest problems on eBay is forbidding people trying to sell Japanese body parts that were once taken as "souvenirs" by U.S. troops. These range from pickled ears to teeth, jaw bones and skulls.

Such a mentality was and can only be the result of an incessant mass indoctrination campaign, and that is exactly what the Office of War Information carried out. As open conflict was winding down, a war correspondent by the name of Edward L. Jones wrote in the February 1946 edition of *Atlantic Monthly*.

> "What kind of war do civilians suppose we fought, anyway? We shot prisoners in cold blood, wiped out hospitals, strafed lifeboats, killed or mistreated enemy civilians,

finished off the enemy wounded, tossed the dying into a hole with the dead, and in the Pacific boiled the flesh off enemy skulls to make table ornaments for sweethearts, or carved their bones into letter openers."

With Goebbels serving as the indirect mentor of the OWI, it is prudent to cite his most famous quotation:

"If you tell a lie big enough and keep repeating it, people will eventually come to believe it. The lie can be maintained only for such time as the State can shield the people from the political, economic and/or military consequences of the lie. It thus becomes vitally important for the State to use all of its powers to repress dissent, for the truth is the mortal enemy of the lie, and thus by extension, the truth is the greatest enemy of the State."

The purpose of this book is to serve as a dagger into the heart of the lies that have made slaves of an entire generation and also all their descendants. Accordingly, another quote from Goebbels is relevant here.

"The essence of propaganda consists in winning people over to an idea so sincerely, so vitally, that in the end they succumb to it utterly and can never escape from it."

You can escape, but first you have to recognize the truth

CHAPTER FOURTEEN

MIDWAY

At the Battle of Los Angeles, Emperor Hirohito accomplished what was his equivalent of what the Russians did with Stalingrad or Leningrad. He wiped out several American divisions just by flying a few super-dirigibles over Los Angeles. The purpose of the Battle of Los Angeles was to flex muscle and demonstrate to the Americans what the armed forces of the Emperor were capable of. While it is glaringly obvious that the Emperor could have unleashed tremendous biocidal weapons, he did not want a biocidal war but rather open markets and a free Japan that was neither colonized nor restricted economically. He did not, however, predict that the American Army would self-immolate as a result of extreme panic. The Office of War Information kept the Battle of Los Angeles a secret, and it is a secret which is mostly very much intact to this day.

The Americans had now suffered two crushing defeats, the first being Pearl Harbor. America responded with what amounted to be a very lame response by having Jimmy Doolittle lead a bomb raid of Tokyo in April of 1942. While it did no serious damage at all in terms of military strategy, and Doolittle thought he would be court-martialed for his poor performance, Roosevelt promoted him to General and gave him the Congressional Medal of Honor. The OWI cited the bomb raid as a major success and spun it into a major American victory in terms of strategy. It was not, and you can verify this historically. A motion picture entitled *Thirty Seconds Over Tokyo* starring Spencer Tracy was made to commemorate the raid and to portray Doolittle as a hero.

The Battle of the Coral Sea followed less than a month later, a battle which was a decisive victory for the Japanese although that has been spun into a "strategic" victory for the Allies. Both the Doolittle Raid and the Battle of the Coral Sea are spun to make you think the Japanese panicked so as to rush too quickly in order to reach their next objective: the Midway Islands, 1,300 miles northwest of Honolulu. The Midway Islands are really nothing more than a rather modest coral atoll with a total land area of 2.4 square miles that was used as an American air base.

OWI propaganda, and very effective propaganda at that, has literally brainwashed America and most of the world into thinking that the Battle of Midway and the defeat of Admiral Yamamoto was a major turning point in the war because the Japanese lost their first-line aircraft carriers and therefore lost the ability to wage an offensive war in the Pacific. While this sounds very plausible to a person without an understanding of the military strategy at work, nothing could be further from the truth.

Although it has been completely lost to historians, Emperor Hirohito achieved two major objectives as a result of the Battle of Midway. The first was the invasion of the Aleutian Islands, a one thousand mile chain of islands stretching out from the Alaskan mainland. Midway was of minimal importance whereas the Aleutian Islands are comparatively close to Japan and it was an ideal location for the Americans to send bombers to Japan. While the Americans were focusing on Midway, the Japanese successfully invaded the Aleutians. This proved to be of great vexation to the Americans, and you can look at the geography of the situation on a map. The Americans were forced to redeploy a full one-third of their personnel in the Pacific towards attempting to reclaim the Aleutians. More decisively, it also forced the Americans to engage in an extended campaign to try and take islands in the Pacific in order to secure landing strips to bomb the Japanese mainland that otherwise would not have been necessary. This meant taking island after island, an exhausting and fruitless strategy that, in the end, never succeeded.

The OWI has also curtailed the factual history of the Japanese presence in the Aleutians and their actual time spent there, but this is propaganda. If the Americans had true control of the Aleutians, they would have been able to bomb the hell out of Japan far earlier in the war. This process of disinformation also applies to almost every aspect of the war.

With regard to the Battle of Midway Island, there was another key objective obtained by Hirohito, and that was the defeat of Admiral Yamamoto which eventually led to his assassination at the hands of the Emperor. A very colorful, charismatic and popular character, Yamamoto was viewed by Hirohito as highly erratic in his performance as a military leader and a danger to the Empire. With a roguish image, Yamamoto was loved by the geishas, common sailors and most everyone else, so removing him from his position would have been a disastrous public relations maneuver. Instead, Hirohito slated him for the Battle of Midway, knowing damn well that the Americans had broken the Japanese codes. Although these broken codes are hailed as the reason for the Americans' success at Midway, it was all a ruse. Hirohito knew the Americans had the codes and wanted them to focus on Midway rather than the defense of the Aleutians. As for, Yamamoto, he made several crucial decisions that resulted in the Americans inflicting far more casualties than was expected by either side. In fact, the Americans

MIDWAY

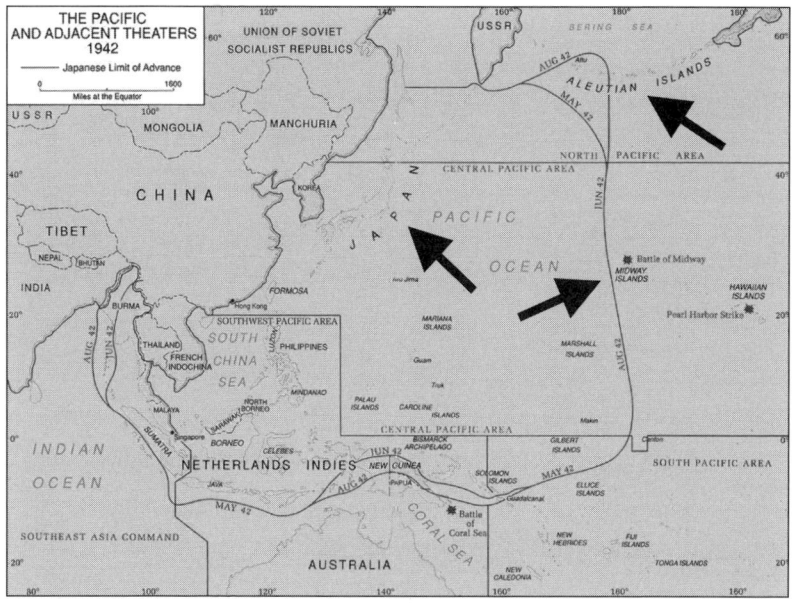

PACIFIC THEATER SHOWING JAPAN, ALEUTIANS AND MIDWAY

have hailed their "success" at Midway as a miracle because what they eventually achieved was so improbable. Yamamoto's incompetence was the sole reason for the miracle.

    Hirohito subsequently dispatched Yamamoto to inspect the front lines in the South Pacific, deliberately arranging for his specific flight itinerary to be delineated so that it could be transmitted in order that the Americans could intercept it and break the code. Despite urgings from the local commanders not to fly, Yamamoto could not disregard Imperial mandate, and his plane was intercepted by fighters and shot down. He was killed, and subsequently duly honored as a hero. The direct hit was ordered by Admiral Nimitz; and once again, the Americans had taken Hirohito's bait.

THE ROSWELL DECEPTION

CHAPTER FIFTEEN

Fu-Go Balloon Bombs

Exploiting the purported victory of the Battle of Midway, the OWI spun the idea that the Japanese could not wage an offensive war against America. This was an absurd statement after stunning defeats in Hawaii, the Aleutians and Los Angeles. These were major aggressive incursions on U.S. territory itself. While the OWI was claiming that Japanese ships and submarines could not come to the United States, Japan was sending ships, subs, and also the first intercontinental weapon ever delivered, the Fu-Go or "balloon bomb". Due to the powerful speed of the Kuroshio Current, the Japanese could have individually come to the U.S. in fiberglass bath tubs if they had wanted to and indeed have demonstrably done so.

The Japanese balloon bombs, known as Fu-Go, were made out of rice paper. They were glued together with rice candy that was literally edible. They cost very little comparatively, each one costing about $50, and the Americans could not possibly shoot them down due to the

Illustration of a fleet of Fu-Go balloon bombs

sheer numbers of them coming across the Pacific. Despite whatever you might read in American press reports, Wikipedia or any other embedded publications, there were over 120,000 Japanese balloon bombs that generated at least 100,000 fires in North America, most of them in the Pacific Northwest. This is way higher than the common reports stating 9,300 balloons and six people dying.

Just as the super dirigibles in the Battle of Los Angeles did not contain biocidal weapons, neither did the Fu-Go balloons, but they did have incendiary bombs which generated dozens upon dozens of firestorms. To put this in perspective, understand that a firestorm is not a mere forest fire or the like. A firestorm, by definition, creates a powerful updraft which causes very strong in-rushing winds to develop in the surrounding area. It is something so huge that it begins to induce its own kind of internal combustion and feed itself, like what happened in the firebombing of Dresden.

To give yet further perspective on how extensive this was, the U.S. Government admitted in 1939, the same year that war erupted in Central Europe, that 503 million acres of U.S. land remained uncharted in the United States. Even today, there are about 383 million acres of Western federal land that has never been surveyed. That is 17% of the entire United States, and to that can be added another 50 million acres that were surveyed inadequately about one hundred years ago. That's another 2.2% right there. You can compare that to a 1990 census revealing that 143 counties in 16 of the Western states have less than two people per square mile. That's less that 273,000 people in less than 950,000 square miles. That is 25% of the entire area of the United States that we know very little about.

The Emperor, however, was not igniting firestorms for purposes of gratuitous destruction. And while these fires had a devastating psychological impact upon the Army personnel who knew about them, Hirohito was burning these forests to measure the wind currents by using planes that would take off from catapults attached to submarines. The Japanese I-400 submarine (see photograph on page 146) was the largest submarine ever made until the advent of the nuclear submarine, and it was equipped with catapults for multiple airplanes, but even the standard workhorse Japanese submarine had at least a single sea plane it could launch for reconnaissance purposes. The result was that there were Japanese pilots flying over the United States all the time, reconnoitering the smoke from the fires so they could map the wind currents. For anyone who does not believe that, they do not understand how accurately these wind currents were mapped. There is a major example of this; and while it had major ramifications for the war, it is commonly

DIAGRAMS OF FU-GO BALLOON BOMBS

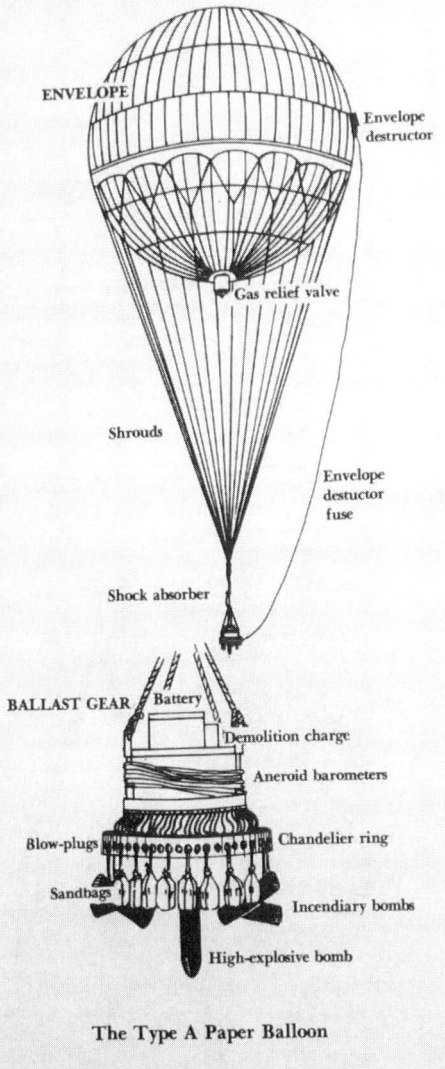

The Type A Paper Balloon

model, which was experimental, there were several methods for dropping ballast sand. Incident No. 1 had 8 bags containing 10.5 kg. of sand each.

BALLAST GEAR: The ballast systems used for the paper and rubber balloons were completely different. In the paper balloon,

THE ROSWELL DECEPTION

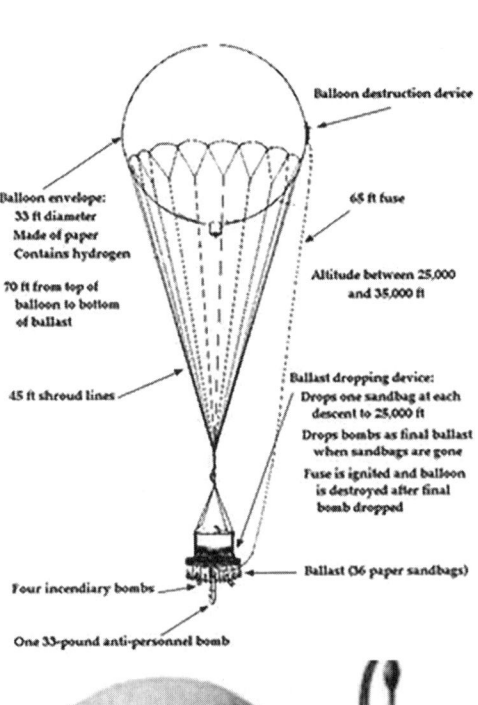

Balloon destruction device
Balloon envelope:
33 ft diameter
Made of paper
Contains hydrogen

65 ft fuse

70 ft from top of balloon to bottom of ballast

Altitude between 25,000 and 35,000 ft

45 ft shroud lines

Ballast dropping device:
Drops one sandbag at each descent to 25,000 ft
Drops bombs as final ballast when sandbags are gone
Fuse is ignited and balloon is destroyed after final bomb dropped

Four incendiary bombs

Ballast (36 paper sandbags)

One 33-pound anti-personnel bomb

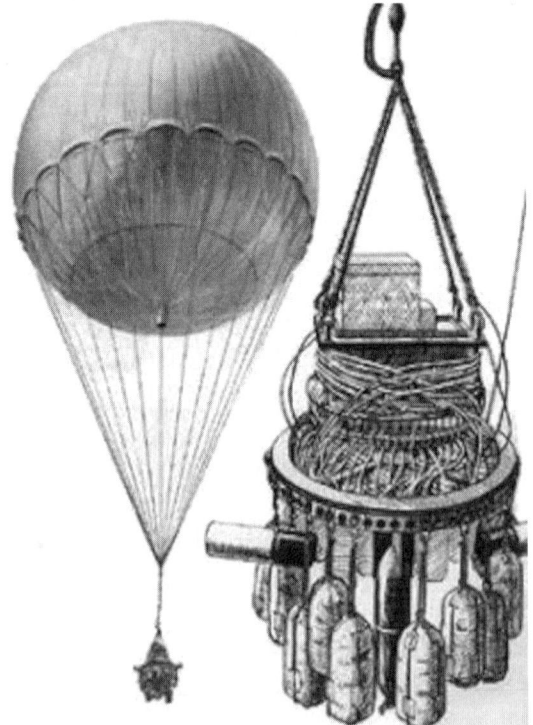

THE FU-GO BALLOON BOMBS

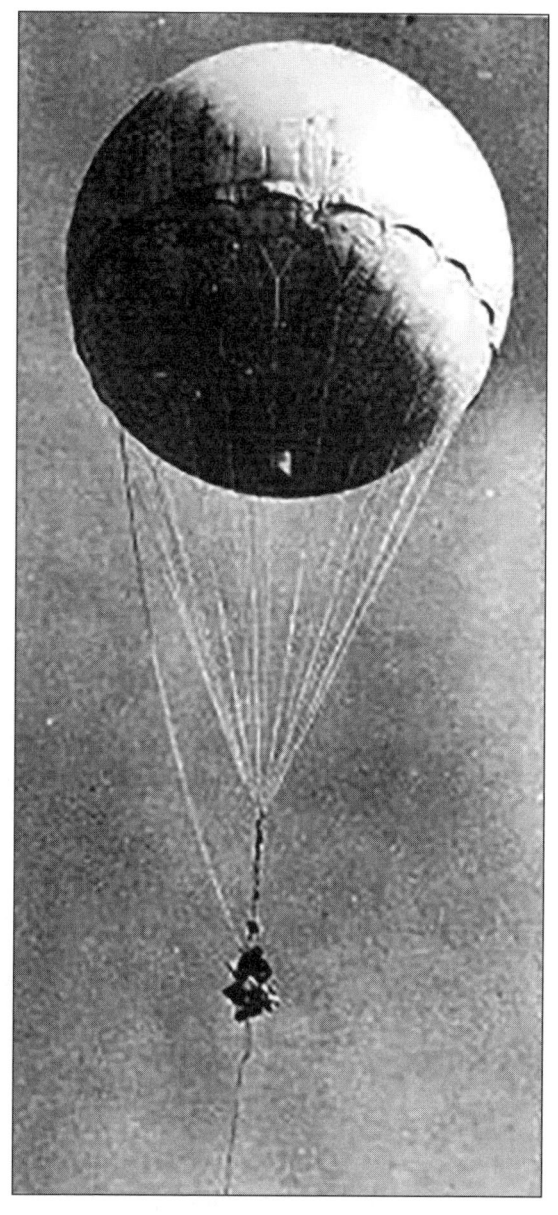

ABOVE IS AN ACTUAL PHOTO OF A FU-GO BALLOON BOMB

PHOTOGRAH OF JAPANESE I-400 SUPERSUBMARINE

ILLUSTRATION OF JAPANESE I-400 SUPERSUBMARINE

and inaccurately portrayed by the OWI and its outlets to have been an "accident".

On March 9th through the 10th of 1945, 334 B-29s belonging to the United States dropped 2,000 tons of firebombs on Tokyo. With precise knowledge of the wind currents, Emperor Hirohito responded with a single wind-current-directed balloon bomb that floated in low across the Yakima Valley through a massive rain front, to its intended target no less, to strike down the main transmission line carrying power from Bonneville Dam to Hanford Works in Washington state, the site involved with the generation of plutonium for the Manhattan Project, which itself was headquartered at Los Alamos in New Mexico. Even though Hanford was a top secret site, Hirohito knew about it and its importance with regard to the atomic bomb. Hanford remained a top secret site until the 1990s when it was condemned as one of the most

THE FU-GO BALLOON BOMBS

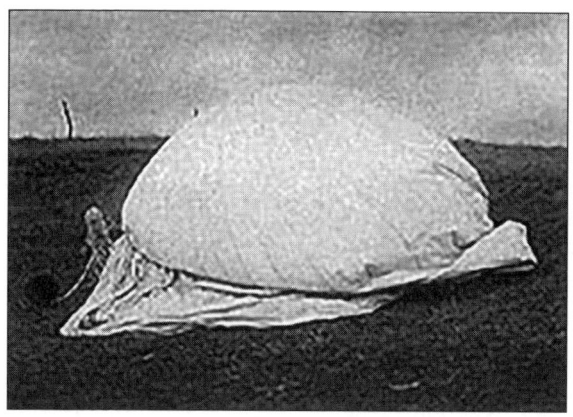

FU-GO BALLOON AFTER COMPLETING ITS MISSION

radioactive wastelands on the face of the Earth. The strike of this single balloon bomb, however, was so devastating that it forced a suspension of the entire Manhattan Project from March 10th through March 12th. The Manhattan Project went into shutdown status because it took the Hanford personnel three days to bring the piles back up to operational capacity. And further, it showed a total security compromise of the entire Manhattan Project. That is how accurately Hirohito had not only mapped the wind currents of the United States but also the topography beneath them.

As the Americans were developing their nuclear program, the Japanese also launched balloon trains over the U.S. in order to monitor the radiation that was generated by any potential atomic tests. This was in addition to the Fu-Go balloon bombs. Through this surveillance, they knew when the U.S. had tested an atomic bomb at the Trinity site, and it is by reason of this that they were forewarned of the American nuclear capacity and prepared for it.

It is also noteworthy that Hirohito was also keenly aware, based on the work of his physicists, that if they were to intercept any atomic bomb being flown in to be dropped over Japan, it would generate an atomic pulse in the air that would potentially fry the entire Japanese national grid. It was for this reason that Hirohito refused to allow the interdiction of any single bombers that were without a massive escort because a fly-alone bomber was almost certainly going to be carrying nuclear weapons. Had Hirohito ordered an interdiction of an atomic bomber, it could result in Japan being denied all electrical power, and he could not afford that potentiality. That is why he had to allow the atomic bombs to drop.

A lot of people do not understand that the atomic bombers themselves, the *Enola Gay* and *Bock's Car*, functioned on vacuum tube era technology. If they were truly modern jets full of circuit boards and integrated technology, like the modern unshielded type of passenger jet, if they were to carry and drop such nuclear ordnance, the resulting electromagnetic pulse would cause them to drop out of the sky into the very mushroom cloud they had created. The only reason that did not happen to those two flying super fortresses was because they were running on vacuum tube era technology. Otherwise, their instrumentation would have been fried.

While the Office of War Information could not possibly hide the fact that the Japanese were sending incendiary balloons en masse, they were able to suppress the fact that these balloon bombs were also primed to carry and disperse bacteriological weapons. As most of the firestorms were in relatively unpopulated regions, it was fairly easy to suppress the true threat of what was happening. The fires themselves, however, were a very real threat that the Army took seriously and they created a special new battalion to extinguish them.

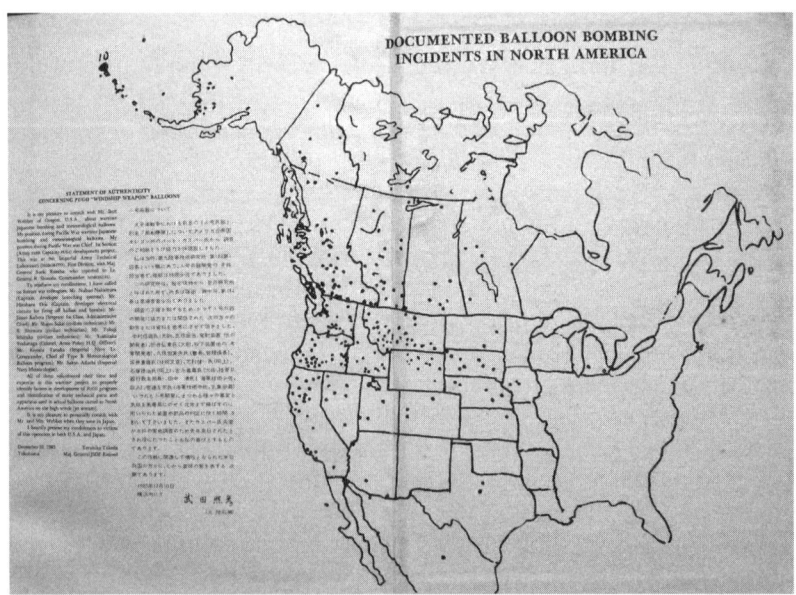

ABOVE IS A MAP OF BALLOON BOMBING INCIDENTS THAT ARE ACKNOWLEDGED BY THE GOVERNMENT. THERE WERE FAR MORE FU-GO BALLOON BOMBS THAT CAUGHT FIRE AND DID DAMAGE.

CHAPTER SIXTEEN

THE TRIPLE NICKLES BATTALION

As the onslaught of Japanese Fu-Go balloons became overwhelming, the Army went to Dr. Robert Goddard, the pioneer rocket scientist who had moved his rocket lab to Roswell, New Mexico. Asking the scientist how they could possibly use his rockets to defeat the Fu-Go balloons, Goddard informed them that the Fu-Go cost very little to make whereas his rockets cost hundreds of dollars. There were also thousands upon thousands of incoming balloons. Aside from the cost, shooting them down with rockets would make for an embarrassingly small kill ratio. While the Americans were already shooting down balloons with planes, it was still an impossible task. It was thousands against a few shoot downs. It was not a game the Americans could win.

ROBERT GODDARD

To counter the firestorms, the Americans organized an all black unit of African Americans and staffed their officer positions with other African Americans. This was the first time in all American history that such an innovation was ever allowed. A very elite unit, they were designated as the "Triple Nickles", the 555th Parachute Infantry Battalion. This was the first such organization in American history, a military smokejumper unit, and they were assigned to extinguish fires in the Pacific Northwest; more specifically, the fires set off by the Fu-Go balloons. As the Government thought it better to keep the people ignorant in order to keep them adequately responding to their propaganda, they exclusively chose African Americans to man this unit, the reason being that the race barrier was so strong that these African American soldiers and officers would basically never be speaking in any normal situation to the majority of the American public. Consequently, most Americans never realized that they were ever under constant direct attack. This cultivated ignorance of the actual circumstances has been maintained to this day.

The 555th was nicknamed the "Triple Nickles" because of its numerical designation (555) and the selection of 17 of the original 20-member "colored test platoon" from the 92nd Infantry (Buffalo) Division. This was the origin of the term "Buffalo Nickles". You will notice, however, that the spelling of "nickles" derives from Old English, a deliberate misdirection by the War Department in order that the history of these paratroopers could remain hidden in the archives. Filing their history in the old search retrieval systems with the Dewey Decimal System of library shelving was enough to throw off any researcher. Nobody was able to research the Triple Nickles for decades because of this old retrieval hand system where an "incorrect" spelling could remain buried. As everyone searching for the history of these very brave warriors were using the modern English spelling of "nickels", they could not find anything. That was but one tactic used to keep this secret.

The patches of these African American soldiers who served in the Triple Nickles was a black panther on top of an open white parachute. Assigned to put out fires, the end result was that the Triple Nickles became so well trained that they were able to put out what were actual firestorms. These specialists were sent to attack these firestorms in the Pacific Northwest, and ultimately, they made about 36 fire calls which included about 1,200 individual jumps, a tremendous endeavor by any standard. Putting out 36 firestorms is an enormous feat in and of itself. Bear in mind, these were actual firestorms and not mere forest fires. The enormity of their accomplishments was truly heroic and cannot be over emphasized.

PATCH FOR TRIPLE NICKLES 555TH PARACHUTE INFANTRY

SMOKE JUMPERS OF THE TRIPLE NICKLES

These men gained so much practical experience, they called the program Jet Black because they likened them to human jet fighters, a direct comparison to the Third Reich's Messerschmidt jet fighter wing which had stacked up the highest kill ratios of any aerial combat unit. The kill ratio of the Messerschmidt has never been equalled. The end result, however, was that these German jet fighters could still not make headway against the American bombers because the Americans were, quite literally, turning out a bomber every minute on the assembly line; so, if each German jet fighter was shooting down five bombers each, the Americans were throwing in hundreds. By sheer brute force of numbers, the Americans were overwhelming the Germans in terms of a bombing campaign. This is exactly what the Japanese were doing with balloon bombs. The end result was that you had the African American paratroopers fighting the fires they produced, but there were not enough Jet Black fighters to stop the barrage of an ever increasing wave of Japanese balloon bombs.

Although they did not have time to train any new firefighters, the Triple Nickles program was expanded in a very sinister way, and in a way that the regular 555th Parachute Infantry Battalion (PIB) never even knew about. That involved African American conscientious objectors. An extension of the Triple Nickles program, these African American conscientious objectors were part of the Jet Black Program and were totally off the books and consequently never heard of. As they were not soldiers per se, they were part of the Civilian Conservation Corps (CCC).

The CCC was originally an army of labor organized by FDR to find people employment. Like everything else in the Roosevelt administration, the CCC was racially segregated. Accordingly, there were white conscientious objectors who were also smokejumpers and fought fires as public civilian servants under the jurisdiction of the Forestry Service. These white smoke jumpers of the CPS or Civilian Public Service proved that they were physically fit enough to become smokejumpers, a job which is one of the most demanding in the world. Instead of fighting in the war, they could prove they were not cowardly by fighting fires on the American home front. Only half a thousand men, these white smokejumpers made fire jumps; and after the war, a museum was made to honor their service. To the contrary, the Triple Nickles were buried in obscurity. When you learn what happened with the black conscientious objectors, you will better understand why the history of the Triple Nickles was so obfuscated. So much of this has to do with the heritage of a very remarkable African American named Herman Perry, his surname having been acquired by reason of his descending from a slave of Commodore Matthew Perry, the man who invaded Japan and made the first strike in a clash between two cultures that would eventually lead to World War II.

CHAPTER SEVENTEEN

BLACK LIVES DIDN'T MATTER

Most African American communities were ambivalent throughout the war. Many among them were African American conscientious objectors. One of them, Robert Leonard Jordan, had worked with the Japanese Merchant Marines and said "the Japanese are our colored brothers", encouraging his people not to fight on behalf of white America as they were still being enslaved by them. All of this came to a head and was driven home to African Americans when Herman Perry, an African American soldier building the Burma Road for Americans, was almost whipped to death by his white supervisor. Pulling out his supervisor's gun, he killed him with it and became the subject of the largest manhunt in human history.

In order to build a road crossing the China-Burma-India (CBI) borderlines so as to send supplies to the Chinese during World War II, a huge construction project was undertaken. The soldiers, however,

HERMAN PERRY

were deceived about their actual assignment. Told they were going to build air strips, the 849th Engineer Aviation Battalion was subjected to harsh tropical conditions, unduly long hours and extreme hardship. The workers of the 849th were all black, and there were about fifty white officers.

While there are different sanitized versions, none of them conveyed the truth: that Perry was whipped to within an inch of his life for alleged negligence of duty. Treated unfairly and subjected to injustice on all too many levels, Perry was driven to kill his superior and escaped, thus unleashing the biggest manhunt in the history of the American Army. He was, quite literally, an enslaved African-American who fought back.

Perry found a home with head-hunters in the jungle and even found a native wife. As word spread of this black man amongst the natives, he became known back home to the black community as a folk hero who was referred to as "The King of the Ledo Road". The Army ferreted him out, ambushed him and shot him as he escaped from his hut. Despite a severe wound in his chest, he began to recover in the hospital although he was very drugged up. During this period, the Army forced him to sign a confession and this sealed his fate at his court martial. Although his defense attorney never brought up the injustice he had suffered, he did appeal and Perry's execution was delayed. During this time, Perry escaped and an even bigger manhunt was unleashed. After suffering multiple wounds, being captured again and making multiple escapes, he was finally hanged.

As news of Perry circulated throughout the world, it dampened the spirit of African-Americans as far as participating in the crusade against the Japanese. As was alluded to above, this story was an emphatic inspiration for African-Americans to conscientiously object to the war.

It cannot be over emphasized how racist and segregated the U.S. Army was at this point in history. In the wake of Herman Perry and what was an extension of the 555th Triple Nickles Battalion, African-American conscientious objectors were mobilized and were given what were called "black parachutes". They were not black in color but were only for black men. They were purposely designed to be so small that, when they opened up, the drag force would be so tremendous that it would break every single bone in their body. As the parachutes would fall to the ground, the jumpers would be hanging limp, falling into the Japanese balloon bomb fires as they were being burned alive. The U.S. Army killed thousands of African-Americans like this, and their bones are still out there in the Western states where no one can find them. This was one major reason, but not the only one, why the U.S. Army deliberately obfuscated the history of the Triple Nickles Battalion.

CHAPTER EIGHTEEN

OKINAWA — OPERATIONS DETACHMENT AND DOWNFALL

As I have already delineated, the war in the Pacific was very tough for the Americans. After being attacked at home, their only success was to manage turning a costly defeat in the Aleutians into a public relations victory vis-a-vis Midway.

In February of 1945, the Americans initiated Operation Detachment which was the invasion of Iwo Jima, a small island of eight square miles located 750 miles south of Tokyo, in what proved to be the bloodiest battle of the war with over 26,000 American casualties. This was one third of all Marine Corps casualties during the entire war.

Although the Battle of Iwo Jima has been highly glamorized with propaganda and the iconic image of the raising of the American flag, it has been severely criticized as being a Pyrrhic victory by military historians. Although the strategy was to use it as an air base for providing fighter escorts on bombing raids, only ten such escort missions ever occurred, and this did not justify the extreme loss of life. Although Iwo Jima was only a tiny island and very far from the main islands of Japan, it was a genuine piece of the Japanese Empire, but it was impractical as far as serving their intended invasion of the main islands.

The next strategy of the Americans was Operation Downfall, the proposed Allied plan for the invasion of the Japanese home Islands. Before they could pull that off, however, they chose to invade Okinawa as a staging area by which to carry out the invasion of Japan itself. A considerable amount of confusion and misdirection has arisen surrounding the invasion of Okinawa. The main island as well as the capital of what are known as the Ryuku Islands, Okinawa is only 466 square miles in area. It is a separate nationality and a separate ethnicity from the Japanese. A Sinitic people, like the Chinese or Vietnamese, they fly their own flag. For their part, the Americans were occupying a different country in Asia yet often claim they invaded Japan, thus deserving all the mockery that has been inflicted upon them due this delusional assertion. But, what really happened in Okinawa?

First, it is important to understand that the Okinawan, who are renown for their longevity as a people, eat pork, unlike the Japanese. The Japanese, who are vegetarians who also eat fish, were comparatively malnourished as they ate no meat protein. They were very short. As the Okinawan raised pigs and had a considerable pork diet, the American strategy in invading Okinawa was to track down and kill every pig on the island. By doing this, the Okinawan would starve to death as a race by the time winter would come around. This was intended to be complete genocide, and bear in mind, Okinawan are not Japanese people. They were, however, allies of the Japanese, the latter making every attempt to help them in defending the island.

The Americans invaded the tiny island of Okinawa on April 1st of 1945. Much larger than D-Day, it was the largest invasion fleet ever assembled in history. Although they were able to take the island, it was yet another disaster for the Americans because virtually the entire American Marine Corps was lost. With the bulk of veterans comprising the experienced Marine divisions lost, there was effectively no one to train new marines. What was even more intimidating to the Americans than their collective losses was they way that they lost them. I am referring here to the Kamikazes. They were first recognized as such in October of 1944 upon MacArthur's return invasion of the Philippines when a U.S. escort carrier was sunk and three others were severely damaged off the coast of Leyte. There was no question these were suicide pilots, and what was worse was that the attacks only increased throughout the rest of the year. As many as three Kamikaze planes would target a particular ship, and besides those, there were fleets of them. On the Philippines alone, 400 suicide craft sank 16 ships and damaged 87 others.

The Kamikazes were so terrorizing that very strict regulations were instituted and enforced to censor any mention of them in letters outside of the immediate theater. To the troops on the ground and on the ships, these were referred to as suicide attacks. They did not refer to them as Kamikazes until English broadcasts by the Japanese through Radio Tokyo and Radio Saigon called them Kamikazes, and this is how the word became a part of the American lexicon. The Americans and the Allies wanted to keep these attacks secret because it would not only discourage their own troops and public, their incredible success would only encourage the enemy.

The intensity and effectiveness of the Kamikazes only increased when the Americans invaded Okinawa. Four times as many suicide aircraft were used, sinking another 16 ships and damaging 181. In addition, the Japanese also used "Baka" or piloted flying bombs that were released from "Betty" bombers miles from the target. These included solid-propellant rockets in the rear of the craft with the pilot acting as the human guidance system. The psychological impact upon the American was so devastating that it could not have helped but contribute to their savage attack on the Okinawan people.[*]

---

[*] ref. *The Incredible "Weapon": The Suicide Attack*, by James P. Gallagher.

OKINAWA

**ABOVE IS AN ACTUAL ROCKET BOMBER
DESIGNED FOR KAMIKAZE ATTACKS.**

While the Okinawan were not Japanese themselves, they were extremely resistant to the American attack and were willing to fight to the death. The barbaric American onslaught included taking old people and burning them alive for fun and taking the children and roasting them on spits and eating them. This only made the entire population resist even more. The Americans raped, killed and tortured people so savagely that many Okinawan people killed themselves to avoid gang rape and torture to the death, including being burned alive. The American behavior was so outrageous that a vast number of Okinawan people chose to kill themselves. In fact, more Okinawan died by suicide than there were Japanese killed in Hiroshima and Nagasaki put together.

The abominable American behavior on the field of battle made it a fight to the death that was even more vehement than it would have been otherwise. If the Unites States invaded Japan, it was going to be just like Okinawa every inch of the way, only worse, as nobody was going to die without fighting to the death. This is why it was so hard to invade Japan itself, especially with no Marine Corps.

At that point, the Americans were hard pressed. A land invasion was obviously doomed. If they were put through all that devastation from a tiny little island, what havoc would they wreak upon themselves if they invaded Japan itself? Operation Downfall was cancelled. Instead, they opted for their last desperate hope: the atomic bomb.

ABOVE IS A DIAGRAM OF A JAPANESE ROCKET BOMBER.

OKINAWA

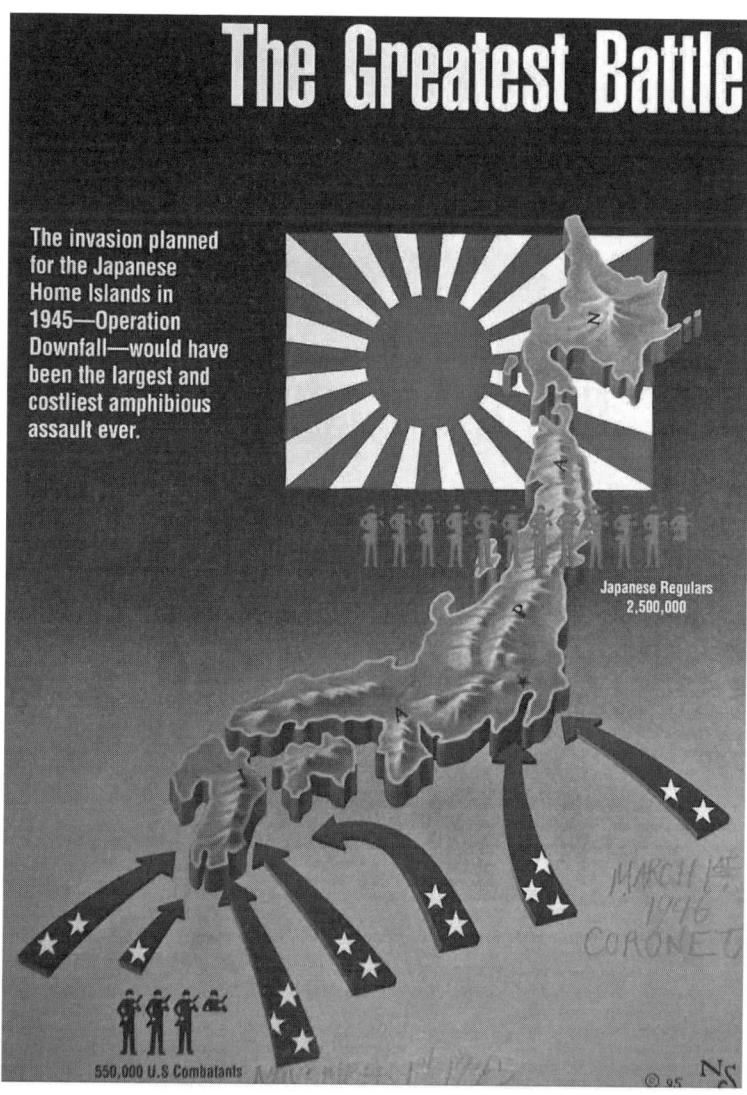

THE ABOVE GRAPHIC FROM A MARCH 1, 1946 ISSUE OF *CORONET* MAGAZINE DEMONSTRATES THE FUTILE ODDS AGAINST THE AMERICANS HAD THEY ATTEMPTED TO INVADE THE JAPANESE MAIN ISLANDS.

CHAPTER NINETEEN

ATOMIC WARFARE

The greatest misconception that has been perpetrated upon the American public by the Office of War Information is that the Americans created the atomic bomb; and further, that they were the first to deploy it, thus resulting in frightening the Axis powers into unconditional surrender.

In point of fact, the Third Reich was the first power to develop and deploy the atomic bomb. This occurred in October of 1944 in Courland, a part of Latvia, where it was used as a deterrent to the Russians who were trying to invade the Baltic nations of Estonia, Latvia and Lithuania. As the Germans considered these to be Germanic nations that had been part of the Order of Teutonic Knights for thousands of years, they wanted to rescue as many of these "German nationals" as they could. In his Polish war crimes affidavit, SS General Jakob Sporrenberg told about his help with developing the atomic bomb and its development and eventual deployment at Courland. There was also an affidavit released during the Clinton administration which concerned the interrogation of a Luftwaffe test pilot who provided evidence that the Third Reich tested an atomic bomb, evidently via a very large primitive cruise missile. It ignited a mushroom cloud on the Baltic coast that was a full kilometer in diameter with continuous internalization of combustion generating electrical interference with instruments over an area extending from Berlin to Great Britain.

Despite the initial success of the German atomic bomb, medical doctors explained to Hitler that radiation caused by that bomb inflicted cellular level damage. This was the first time in history that an atomic weapon had been deployed, and through its deployment, the German medical community was able to identify the fact that radiation does not behave like a virus and that a cell cannot fight off the radiation with antibodies. Unlike a disease, radiation does not fight by the rules: the body cannot defend itself against the radiation nor build up an immunity. Adolph Hitler was obsessed with biological analogies. He viewed anti-partisan warfare as an immunization process and actually used that analogy multiple times. When he was confronted with this reality of atomic weapons, Hitler felt that any further deployment on the continent of Europe would be a threat to the genetic viability of Aryan man. He swore that, no matter what happened, he would never again deploy atomic weapons on the continent of Europe. This did not apply to attacks elsewhere.

Hitler authorized Operation Seawolf, commanded by Admiral Karl Dönitz, which included the launching of the Fenris Wolf Pack consisting of six über U-boats en route to North America that were capable of delivering very large rockets containing nuclear warheads. Because of this, the Commander-in-Chief of the Atlantic Command

Figure 9. I-Gō Radio-Guided missile (I-gō rocket bomb). Courtesy of the National Air and Space Museum, Smithsonian Institution (SI 73-10612).

Figure 10. I-Gō Radio-Guided missile (I-gō rocket bomb) loaded under Ki-48 bomber. Courtesy of the National Air and Space Museum, Smithsonian Institution (SI 73-10615).

THE TOP PHOTO IS AN EARLY JAPANESE CRUISE MISSILE DESIGNED TO DELIVER NUCLEAR ORDNANCE. THE BOTTOM SHOWS THE IDENTICAL MISSILE LOADED ONTO AN AIRPLANE.

(CINCLANT) warned the American public on national radio and through newspapers that they had better prepare for an atomic attack from the Third Reich. Based on what they had found out from their intelligence units on continental Europe, the British had warned the Americans and Canadians of an atomic attack on April 1, 1945. On April 24th, after nine days of engagement, the last of the U-boats were destroyed by the U.S. Navy. Grand Admiral Karl Dönitz, however, had proved his point: he was capable of a nuclear attack against the United States. The reason why Adolph Hitler gave him the go-ahead for this was because the aforementioned deployment of the atomic bomb at Courland in 1944 had stopped the Soviet forces from advancing in Estonia, a region that remained in German hands even after the fall of Berlin.

At the end of the war, the Americans and the Soviets propagandized that every inch of land in Germany or continental Europe basically had been swept or occupied by either the Soviets or the Americans, but this was not true. Besides Estonia, there were several factions and strongholds of German military resistance that had not surrendered, even when it was reported that they had. In other words, the Germans were in a position to barter. Although they were clearly losing the ground war, they had considerable cards to play, including advanced technology that ranged from computers to rockets and other flying craft. There was also the atomic bomb.

By 1944, the Germans had already begun a massive program to relocate technology, personnel and other resources to the Southern Hemisphere. A land invasion of Germany was inevitable. Much of this exodus strategy had to do with the stubbornness of Franklin Roosevelt.

You might have already read historical accounts of Hitler being ecstatic upon learning of Roosevelt's death in April of 1945, not too long before Hitler had his final moments in the bunker. Historians will often chide Hitler for being so exuberant over the death of a rather feeble president whose presence on the international scene was deemed to be relatively inconsequential. According to documents I was dealing with, President Roosevelt was actually assassinated by Axis agents. More popular theories have been floated that the British or even the communists were behind it, but the most plausible theory is perhaps that his oatmeal was tainted, and that it was done through a domestic servant that was indeed an agent. Without going further into the different aspects of that case, such as the facts that there was no proper medical examination and that the records of FDR's death have been sealed, it is important to understand the motivation for eliminating Roosevelt, particularly from a German perspective.

As Roosevelt had demanded an unconditional surrender from both the Germans and the Japanese, this only prolonged the war. Declaring Hirohito, Hitler and their governments to be war criminals, Roosevelt vowed not to deal with them in any way, shape or form. He wanted them dead. This meant the war could never end other than in the extermination of the German and Japanese races from the face of the Earth. This mandate completely frustrated General Eisenhower who was ordered only to kill, kill, kill. Eisenhower tried to get around this by bringing in the famous band leader, Glenn Miller, who spoke German and was popular in Germany, to conduct secret peace negotiations with the Third Reich. That is how desperate the Americans really were. Glenn Miller died very mysteriously and the Americans organized an entire cover story that he was killed in an aircraft crash. That is not what happened.

Neither Stalin nor Churchill were in favor of this unlimited killing policy, and as a result of this, they have been suspected of killing Roosevelt themselves. Roosevelt's extermination policy, however, bottlenecked the Germans even more, and it very clearly explains Hitler's exuberance over FDR's death. The death of FDR facilitated various talk-down processes which included different negotiations and "deals" that were made. While the repatriation of the Paper Clip scientists is the most popularly recognized, there were also many other issues to be dealt with.

As the hostilities wound down, the Americans could not communicate with the Third Reich government but could only accept cease-fires in the field from the military. This awkward position extended even to the publicly recognized end of the war when General Alfred Jodl and Grand Admiral Karl Dönitz surrendered the German military forces in what amounted to a cease-fire as opposed to either a formal capitulation or surrender by the legally constituted German state.

While the German forces in the field surrendered to the Allies, the German government itself was not spoken to and was simply allowed to leave. The people who were in the dock at the Nuremberg Trials were all military men. Even Albert Speer was considered military because he was in charge of military production.

One of the reasons why Grand Admiral Karl Dönitz of the U-boat arm of the Kriegsmarine was chosen to become the last Führer of the Third Reich after Adolph Hitler was that he was in charge of so many important operations. One of them was Operation Hannibal, the largest evacuation of civilians in the history of humanity. This included relocation to Argentina, then to Antarctica, and eventually, to what was referred to as Unterland.

The most significant of the aforementioned negotiations that took place during this period concerned the American acquisition of the atomic bomb from the Germans. The atom bomb that was dropped

ATOMIC WARFARE

on Hiroshima was actually a German bomb, and it was referred to as "Little Boy" by the Americans. It was a standardized piece of atomic ordnance that was developed by the Axis who had bombs by the dozens. But the Axis, both the Germans and the Japanese, viewed these as operational-level ordnance, neither tactical nor strategic. They were never designed to end a war but rather to break an army in the field. The Germans were the first to develop them. Little Boy was built very specifically to fit inside the fuselage of a medium bomber, the bomber itself serving as the delivery system. In other words, it would be a suicide bomber. The Japanese were very comfortable with this, but the Germans also developed a suicide pilots corps that was willing to utilize this weapon. They never chose to deploy it.

When Harry S. Truman took over from Roosevelt, his first act as President was to recall the aviator Charles Lindbergh from the Pacific immediately and dispatch him to Europe. Charles Augustus Lindbergh spoke German and had been awarded a very high level civilian medal by Adolph Hitler for his sympathies with Germany before and during World War II. Truman assigned Lindbergh to Europe to deal with as many Nazi scientists and technicians as he could in order to bring back the atomic bomb from the Germans.

The Little Boy bomb, which was never tested by the Americans, is distinctly different from the plutonium bomb tested at Trinity, the latter being named Fat Man. It is important to disambiguate these two very different bombs. Little Boy, a uranium bomb, was dropped on Hiroshima on August 6, 1945. Fat Man, a plutonium bomb, was dropped on Nagasaki on August 9, 1945.

While the Americans had a virtual monopoly on plutonium, the Germans and the Japanese had none. Both countries did, however, have an ample supply of uranium, and they had a standardized Axis shell meant to fit into the fuselage of a medium-range bomber. This is what

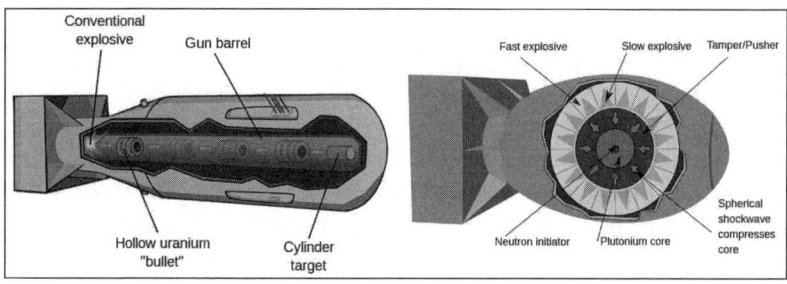

ON THE LEFT IS A DIAGRAM OF "LITTLE BOY",
ON THE RIGHT IS "FAT MAN".

the Americans called Little Boy. By producing a number of these, the Germans and Japanese could threaten the Americans with a medium-range inter-ballistic missile delivery system, a rocket system launched from a U-boat or a Kamikaze plane. Because of this capability, they were able to force multiple concessions from the Americans. When Charles Lindbergh went to continental Europe, he was able to do what Glenn Miller did not do and succeeded in talking the German scientists into giving over a "Little Boy" bomb as part of the concessions that they were dealing with in terms of the Third Reich's government. When this Little Boy bomb was delivered to the United States, it was never tested because the Americans already knew that it worked. They already knew that the Germans had deployed it in October of 1944 at Courland.

For those who would question the veracity of all this, one has to ask, why did the Americans have two separate bomb development systems, one uranium and one plutonium? It does not make any sense, but everybody accepted it after the war as part of the propaganda. The Little Boy bomb was one of the ways that the Germans were able to get concessions. Another was through the Reich's Propaganda Minister, Dr. Paul Joseph Goebbels. From the early days of the war and throughout, he conducted a brilliant propaganda operation in which he claimed that the Third Reich had made contact with extraterrestrials and that they recognized the Third Reich as the legitimate government of planet Earth. This kept the Americans guessing, and as they learned more about his propaganda savvy in this regard, they incorporated it into their own projects such as Operation Mockingbird and eventually created their own alien mythos over the Roswell Incident.

There is no question that the Americans positioned the atomic bomb within a belief system that treated it as what was essentially a religious object. There were multiple reasons for this, but the main reason was money. At that time, there were only two nations on Earth which had an industrial base but did not have an air force: Imperial Japan and the United States. There was no United States Air Force until 1947, and the only way the Americans could get an air force was to sell the lie that their strategic bombing had worked. In reality, however, their strategic bombing had not worked at all. It effectively failed to hinder ever increasing productivity rates under the administration of Albert Speer. Their strategic bombing had also failed to damage the war efforts of the Japanese at all, and this also applies to the dropping of the atomic bombs.

Throughout the majority of the war, 90% of the Japanese army was stationed in or engaged on the Asian mainland, not in the Pacific nor on the home islands of Japan. This also applied when the atomic bombs were dropped on Hiroshima and Nagasaki. The atomic bombs not only did not damage the Japanese military machine, they did not even deter it, but this requires some explanation, much of it concerning the circumstances and strategic outlook as the hostilities peaked.

After the disaster of the Battle of Okinawa, the American high command realized that an invasion of Japan itself would be impossible. The Americans knew the Japanese would fight to the very last man, woman and child, and it would require an even greater deployment of troops, far far more than that which had invaded Okinawa, and that itself was even more than D-day. This would require, at the very least, all of the remaining manpower in Europe being redeployed to Asia. This would require a massive fleet going through the Panama Canal, an area that was already targeted for destruction by the Japanese who had long range submarines fully capable of taking out this route. Hirohito had already deliberately surrendered two of his long range submarines to the United States so they would know exactly what he had. Fully realizing that the Panama Canal would be compromised, the only alternative route for an American invasion was via Cape Horn, and this would leave the Americans hanging at the end of a very long logistical line in the Pacific, and by the time they reached the Japanese home islands, they would be very much left at the end of their rope. Even before the first atomic bomb was dropped on Hiroshima, a special War Department analysis of new Japanese divisions being mobilized reached Army Chief of Staff, George Marshall, and it revealed that, from 1937 to 1943, the Imperial Japanese Army had mobilized an average of eight divisions a year. Bear in mind that the Imperial Japanese Army was the largest employer in Japan with ten percent of the Japanese population working for the army. This was a very big problem for the Americans because, up to that time in the war, the Americans had overwhelmed the Japanese numerically. For example, the Americans were never able to take Makin Island, a key defensive base for the Japanese which was a gateway to attacking Tokyo. In the Battle of Makin Island, the Japanese had 500 troops fighting against an entire American Army division and regimental combat teams with Grants and Shermans (both of which are tanks). By the time Americans got to Okinawa, "the doorstep of Japan", everything changed.

At that point, the Japanese had won the war in China. In 1945, on the very day that George Marshall got the aforementioned report on Japanese resources, he found out the Japanese had mobilized no less than thirty divisions to secure the Chinese mainland in the first seven months of 1945 and that a total of 42 divisions had been newly activated, 23 inside of Japan itself. Further, Japan had the available manpower to generate even more, as many as 65 infantry and 5 armored divisions by the time of the scheduled American invasion of the main island. Up to that point, all of the assets that the Allies had come up against in the Pacific were Japanese marines and naval infantry. There were very few elements of the Japanese army that they had ever encountered, and only in New Guinea. Up until 1945, the Americans had easily outnumbered the Japanese in the field at all times. An invasion of the Japanese home islands was going to reverse the odds by orders of magnitude. Not only would new divisions be mobilized, but the conquering troops on the Asian mainland were also prepared to return to defend the home islands. It was not an optimistic prospect for the Allies at all as they would be outnumbered beyond belief with a tenuous logistical supply line.

In an act that was nothing other than complete desperation, the Americans unleashed Little Boy, the German bomb they had acquired through the negotiations of Charles Lindbergh. Although Little Boy was dropped on Hiroshima on August 6, 1945, the Japanese were less than surprised. They had already long been monitoring America's nuclear efforts as evidenced in the successful attack on Bonneville Dam at the Hanford Works. As for the Americans, they were uncertain if such a nuclear blast would take out their own airplane that was delivering the atomic ordnance. This was a new experience, and they were not prepared to send out a squadron to protect the bomber as those men might simply serve as collateral damage. The Japanese knew it would therefore be a single plane that would carry the bomb, but they knew better than to shoot it down in the air. As alluded to previously, the Emperor's physics advisor, Hideki Yukuwa, who would go on to win the Nobel Prize for Physics in 1949, advised him that a nuclear airburst and the resultant electromagnetic pulse at such an altitude would literally fry Japan's national grid. For this reason, Hirohito allowed it to be dropped. To do otherwise would compromise Japan itself and thereby the Empire. When Little Boy was dropped on Hiroshima, the damage was restricted to civilian damage and did not deter the military at all. As I have stated repeatedly, most of the Japanese Army was stationed on the Asian mainland.

The bomb on Nagasaki, the plutonium bomb known as Fat Man, was a military fiasco from the start and an ongoing embarrassment on behalf of the Americans. The first blunder began with one of the chemists who was dealing with the plutonium. There was only enough of it for one bomb, and all of the plutonium that had ever been produced in the history of the written world was inside a small test tube held by this chemist, Don Mastick. Unknown to him, a chemical reaction was causing pressure to build up inside the vial, and it burst, creating a cloud of plutonium. In a panic, Mastick inhaled and sucked in all the world's plutonium which found its way into his lungs and digestive tract.

In order to retrieve the plutonium to construct the Fat Man bomb, they had to extract it from Mastick's intestines by having him eliminate it through his feces. Bear in mind, I am not making this up nor is it some urban legend. This is historically verifiable, and you can look up Don Mastick. After he eliminated it, they boiled up his feces to retrieve all of the plutonium. The stench that ensued was so unbearable that it literally made all of the Manhattan Project stink. The entire city of Los Alamos was besieged throughout the entire day by the abhorrent odor of his boiling excrement. It was so bad that a petition was drawn up and signed by all of the physicists and everyone else in the city to banish him, and it worked.

When the Americans had invaded Tinian Island, the first item on the agenda of occupation was to fly Don Mastick over as the first civilian colonist so that he could complete his process of defecation over there in order to retrieve every single iota of plutonium, all of which would comprise the incendiary in the triggering mechanism of the bomb itself. This is the reason that the Fat Man bomb arrived disassembled. They first had to retrieve all of the known plutonium on Earth from out of this guy's intestinal tract.

The second major debacle of the Fat Man bomb concerned the misadventures in deploying it. The original target for Fat Man was the Japanese National Arsenal of Kokura, scheduled to be dropped on August 9th by the B-29 bomber *Bock's Car*. Although two other B-29 scout planes reported clear weather, the Office of War Information has contrived to convince the public that it was bad weather that caused *Bock's Car* to abandon its original target and drop Fat Man on Nagasaki as a secondary target. The actual reason was far different.

Bock's Car's flight-engineer was reported to have discovered a major malfunction catalyzed by a puncture from an enemy round which caused the fuel pump to cease functioning and trapped six hundred gallons of highly flammable high-octane aviation fuel in her auxiliary bomb-bay fuel tanks, thus rendering it unavailable for use. In a panicked retreat, Commander Charles W. Sweeney opted to drop the single most powerful weapon in the American inventory over Nagasaki, the most Christianized city in Japan; and, coincidentally, a major munitions manufacturing center.* Nagasaki's geography subsequently played a major role in limiting Fat Man's effectiveness, dissipating eighty percent of the plutonium

---

*John Coster-Mullen Interviewed the late Charles Sweeney in 1995 and recorded Sweeney's recollection of when he met 20th Air Force Commander General Curtis LeMay. It was only a few days after the abortive mission against Japan's national armory, and their "exchange" went as follows, verbatim: "He just looked up at me, didn't return my salute. He simply said: 'You missed the target, didn't you!' I said, 'Yes Sir.' He didn't say, 'Won't you sit down and have a cup of coffee?' That was the same as if you had been condemned to death! My entire purpose as Mission Commander was to get the bomb to the original target."

bomb's power by confining the explosion to the bowl-shaped valley amidst the Nagasaki Hills and thusly preventing casualties, fatalities, and materiel damages from becoming as extensive as those produced by the lower-yielding Little Boy. Based on hospital data, Japan's Asahi Shimbun (*Japanese Times*) estimates of human casualties are 135,000 for Nagasaki, including diseases from the aftereffects.*

Although Nagasaki was a major munitions center, Mitsubishi's production facilities were beneath the city of Nagasaki in underground lead-lined bunkers and remained undamaged. There were well over 2,500 young ladies employed in the underground who survived the deployment of America's most powerful weapon without any awareness that an atomic attack had occurred overhead. Undeterred, the Japanese war machine continued with the on site full-scale assembly of long-range KI-77 bombers, known as "Patsy's", which were capable of reaching New York City from Tokyo. It was concluded (correctly) that any atomic attack on the Japanese national capital of Tokyo would leave the Imperial Bunker(s) equally unscathed.

While there is considerably more intrigue surrounding the "end of the war" in 1945, including Japan's possession of nuclear weapons, the purpose here is to demonstrate how much the Office of War Information has lied to the public. I will offer more information on atomic weapons, but next, I will address the Japanese response to the bombs dropped by the Americans on Hiroshima and Nagasaki.bombs dropped by the Americans on Hiroshima and Nagasaki.

The Spirit of Hiroshima: An Introduction to the Atomic Bomb Tragedy (Pub.: 1999; Hiroshima: Hiroshima Peace Memorial Museum).

ABOVE IS THE TACHIKAWA KI-74 "PATSY" BOMBER, CAPABLE OF FLYING FROM TOKYO TO NEW YORK.

CHAPTER TWENTY

JAPAN'S RESPONSE

Hirohito's response to the atomic bombs was to send three super-dirigibles to America. Launched from the Chinese mainland after the Hiroshima bombing, two reached America in three days, just after the Nagasaki bombing, one having crashed en route. Over one thousand feet in diameter, which is longer than three football fields, these dirigibles followed the wind currents in order to land specifically at Tonopah Army Air Field in Nevada. The Japanese knew that Tonopah Army Air Field was where the Americans developed their most sophisticated aircraft that included their highest level of technological expertise. Tonopah Army Air Field later became known as Area 51.

In response to America's nuclear aggression, Emperor Hirohito deliberately authorized the surrender of these three super-dirigibles. He was not only telling the Americans that he knew their soft spot and could reach it, but he deliberately surrendered the dirigibles so that they could directly witness the deadly biocidal weapons loads they carried. The American brass knew too well the lessons of World War I. It was a war-breaker.

The four feet high Yakuza aeronauts who manned these balloons were captured and treated as prisoners of war by the Americans. The biocidal weapons were immediately taken to Fort Detrick in Maryland, and upon analysis, it was determined that if the contents of one of these dirigibles were distributed evenly across the planet, it could theoretically kill every man, woman, and child on Earth. The Americans had no choice but to pursue peace with Japan, and that is exactly what they did. Nevertheless, they did an excellent public relations job of spinning the news and treaties to make it look like America won the war. There are several aspects that demonstrate this beyond any reasonable doubt, but we have all been so brainwashed or misled that it requires some calm thinking.

The Americans, with only a small percentage of their troops in the Pacific, had suffered over 20,000 casualties in Okinawa alone and were at the end of their rope, having lost their best assault troops. Now, they were confronted with these terrifying super-dirigibles, over a thousand feet in diameter, appearing unimpeded over their most technologically developed outpost with biocidal weapons that had the potential to wipe out the entire human race, let alone the United States itself. There was no other course of action other than to sue for a peace which the

THE ROSWELL DECEPTION

Japanese had been willing to accept for months. The road to peace, however, had a rather extensive history, all of which has been obscured.

THESE PHOTOS SHOW HOW SMALL JAPANESE SOLDIERS COULD BE.

CHAPTER TWENTY-ONE

THE ROAD TO PEACE

The opening month of 1945 (January) saw Emperor Hirohito hold an Imperial Conference during which His Divine Majesty authorized Prime Minister Togo to notify the Allies through General Douglas MacArthur that Japan would accept their terms on one condition – that such declaration "does not comprise any demand which prejudices the prerogatives of His Majesty as a Sovereign ruler".

General Douglas MacArthur himself forwarded this Japanese offer directly to the President by which to conditionally terminate proactive prosecution of hostilities (on the exact terms which the United States was eventually forced to accept seven months into the future). In this overture, one of Hirohito's primary cease-fire demands was the presence of MacArthur in his court to serve as a liaison between His Majesty and the United States Executive. The reason for this is that Hirohito perceived MacArthur to be an egotistically manipulable idiot-savant, a "celebrity-warrior" whose projection of Caesarian authority he had come to admire and who apparently reminded him of his original mentor, General Nogi. As MacArthur was the only prominent United States high officer descendent of multigenerational military lineage, and thereby the closest that American Culture had come to producing that which was comparable to a scion of Samurai nobility, he was someone the Emperor could relate to.

General MacArthur would later himself contend that Japan would have accepted conditional peace before America's commitment to atomic terrorism if the United States had but notified Japan that it would accept a "surrender" that acceded to Emperor Hirohito retaining his position as constitutional leader of Japan. As was already alluded to, this was a condition the U.S. was, in fact, forced to acquiesce to upon suing for peace. The Office of War Information, however, spun the situation to make it sound as if the retention of the Emperor was the result of the insistence of General MacArthur. Hirohito also understood that he could bribe MacArthur and, in fact, did exactly just that, the specifics and true history of which will be addressed in a subsequent chapter.

U.S. leadership cross-confirmed the reality of Japan's overture through Brigadier General Clarke's intercepts of encoded Japanese messages which established that Washington obstinately refused to express any willingness to accept the single condition by which the war could end: that the Emperor would relinquish his title and all leadership of the Empire. The reason the Americans did not accept this overture is that it was well understood that Hirohito himself was the singular supreme source of Japan's strategic success(es) – a military genius of Napoleonic proportions considered far too dangerous of a threat to be allowed to live.

Among others, author Michael D. Gordin, in *Five Days In August: How World War II Became a Nuclear War* (pub. 2007; Princeton University Press), exposits how President Truman had been informed through Swiss, Portuguese and other channels of Japanese peace overtures as early as three months prior to the Hiroshima bombing. Still, the Americans insisted upon an unconditional surrender.

On July 25, 1945, the Allies issued the *Potsdam Declaration*, demanding the unconditional surrender of Japan and threatening complete destruction if this was not done. The *Potsdam Declaration*, however, did not mention the exact way that the Emperor would be treated, but it was clear that they wanted a change to the system of government.

On July 31, 1945, less than a week prior to the U.S. atomic-terror attack on Hiroshima, Hirohito made it clear to Lord Privy Seal Marquis Kido that the Imperial Regalia of Japan had to be defended at all costs. Preservation of the Kokutai (the Japanese system of government) implied not only that of the Imperial institution but also the continuation of God-Emperor Dr. Hirohito's reign. God-Emperor Hirohito, however, was still goading the Soviets into a military response through noncommittal Japanese peace feelers and made no move to change the government's position. The Emperor was instigating an invasion from Russia because he not only knew he could contain it but that such a circumstance would enable him to play the Americans off of the Soviet Union, knowing full well that the United States would be very uncomfortable about a strong and permanent Soviet presence in the Far East. In other words, he was inducing the Americans into a peaceful solution. At the same time, God-Emperor Dr. Hirohito had actively sought the final battle his military was preparing under his direct command because he knew that he would win.

On the morning of August 6, 1945, the B-29 superfortress named *Enola Gay* dropped an atomic bomb on Hiroshima. A short time later, other B-29s began dropping leaflets on Tokyo, translated into Japanese, that stated: "Because your military leaders have rejected the thirteen-part surrender declaration, we have employed our atomic bomb. Before we use this bomb again and again to destroy every resource of the military by which they are prolonging this useless war, petition the Emperor now to end the war."

There was no way that Japanese civilians were in a position to petition Emperor Hirohito to accept the terms of the July 26 *Potsdam Declaration* outlining the Allies' surrender demand, among them the complete disarmament of Japanese forces and the elimination "for all time the authority and influence of those who have deceived and misled the people of Japan into embarking on world conquest". All of those who write historically on this subject are "Americanists". What I find interesting in all this revisionist thesis is that the Japanese have no such agency for world conquest in their own history. Simple history will tell you that it was the other way around. These leaflets, however, revealed a nigh-universally perceived nuanced unreality: that only the Emperor could end the war.

On August 8th in 1945, Japan's Supreme War Council of Dai Nihon (Greater Japan) scheduled a meeting to discuss whether or not to accept the terms of The *Potsdam Declaration* of the United Nations demanding Japan's immediate and unconditional surrender. This meeting was called two days after the American nuclear terrorist attack on Hiroshima.

In the aftermath of America's atomic attack, President Truman announced: "If they do not accept our terms, they may expect a rain of ruin from the air the likes of which has never been seen on this Earth." With both Soviet entry into the greater East Asian war and the aforementioned abortive atomic attack on Japan's national armory at Kokura (destined to be diverted onto the civilian city of Nagasaki) only a matter of hours away, the meeting of the Supreme War Council had to be cancelled because "...one of the Council's members had more important business elsewhere." It is difficult to discern any matter that could have represented "more important business" than consideration of the question of the Japanese nation's immediate and unconditional surrender, save for Imperial coordination of immediate and/or ultimate victory.

It is also paramount to note that between the Hiroshima attack on August 6th, 1945 and the Nagasaki attack on August 9th, 1945, U.S. Chief of Staff General George Catlett Marshall was advised by the Head of Army Intelligence that the atomic bombs were not expected to be of any strategic significance; thus redirecting him back to his original conceptualization of their deployment as a tactical weapon in America's abortive scheme to invade Japan, previously discussed as Operation Downfall.

On the morning of August 9th in 1945, the mission plan for the terrorist attack on Kokura was nearly identical to that of the Hiroshima mission with two B-29's flying an hour ahead as weather scouts.

Observers aboard the weather planes reported: "Target Clear". When Major Charles W. Sweeney's aircraft, the U.S. 393rd Squadron B-29 super-fortress *Bock's Car*, arrived at the assembly point for his flight off the coast of Japan, the third plane, flown by the group's Operations Officer, Lt. Col. James I. Hopkins, Jr., failed to make rendezvous. As a consequence of this failure, *Bock's Car* and her accompanying instrumentation plane circled for forty minutes without locating Hopkins. Already thirty minutes behind schedule, Sweeney decided to fly on without Hopkins. As was covered earlier, the mission to bomb Kokura was a disaster and the atom bomb was dropped on Nagasaki instead, its damage minimal compared to Hiroshima due to the enclosed valley wherein it exploded.

When news of the Nagasaki bombing came on August 9, the Supreme War Direction Council (of Japan) reacted not by moving toward peace but by declaring martial law throughout Japan.

Following the bombing of Nagasaki on the 9th August, the Japanese government had met several times to discuss surrender. There were sharp divisions within their ranks whether they should accept the Allies' ultimatum or fight on. Since the U.S. was reading all of the Japanese secret diplomatic messages, they had an almost real-time understanding of the complexity of the Japanese position.

Col. L.E. Seeman (Aide to Major General Leslie Groves) confirmed that, in the case of an invasion, the blast effect of the bombs would clear the beaches while U.S. soldiers and marines stayed about six miles offshore. The American invaders would land two or three days later.

Lt. Gen. John Hull (Assistant Chief of Staff for Operations and Plans) knew another atomic attack against a city would simply galvanize Japanese "Spiritual Mobilization" further and render surrender in any form impossible and, like the rest of the American High Command, was prepared for a permanent war of racial extermination as not only inevitable but welcome. America's warlords had never dreamed of the American public so willingly giving them such totalitarian powers, and they wallowed in their intent to extend this power-play forever. For Marshall, the only professional soldier ever to win the Nobel Peace Prize, the tactical use of atomic bombs to support the invasion that was primed for Kyushu (the main southern island of Japan) was seen as the only realistic alternative to the continued pounding of Japanese cities; and in this he was simply aping his enemies, and by this, I am referring to the Axis' employment of atomic ordnance in an operational context such as had occurred to great effect at Courland in Estonia.

The day after the bomb was dropped on Nagasaki (August 10th), the Japanese Foreign Ministry transmitted a telegram response to the Allies, apparently offering to accept the terms of the *Potsdam declaration*, but it was most definitely not the full unconditional surrender that the Allies sought but rather "with the understanding that the said declaration does not comprise any demand which prejudices the prerogatives of His Majesty as a sovereign ruler". The Japanese specifically referred to this "understanding" because they had already been offering a "surrender" for months despite the United States leadership's refusal to accept it. There was much scope for interpretation in what the two sides wanted, and it would take five days to resolve the issues at stake.

On August 10, Major General Leslie Groves, military director of the Manhattan Project, sent a memorandum to General of the Army George Marshall, Chief of Staff of the United States Army, in which he wrote that "...the next bomb...should be ready for delivery on the first suitable weather after 17 or 18 August."

On that same day, Marshall endorsed the memo with the comment: "The problem now (August 13th) is whether or not ... to continue dropping them every time one is made and shipped out there or whether to hold them...and then pour them all on in a reasonably short time. Not all in one day, but over a short period. And that also takes into consideration the target that we are after. In other words, should we not concentrate on targets that will be of the greatest assistance to an invasion rather than industry, morale, psychology, and the like? Nearer the tactical use rather than other use."

Despite Marshall's comment, the retention of Hirohito as the ruler of Japan had already been agreed upon in Washington on that tenth day of August.

A full working week prior to a very literal deadline that Hirohito had set,* Allied bombing of Japan was halted on 11th August "because of bad weather", and this situation was confirmed to allow negotiations to continue. Japan then received the Allied reply, including the U.S. insistence that "the authority of the Emperor and the Japanese Government to rule the state shall be subject to the Supreme Commander of the Allied powers who will take such steps as he deems proper to effectuate the surrender terms."

---

*Hirohito, who had been requesting an acceptable peace for years, had now set a final ultimatum and that was for them to agree by the date of August 15th as this was the anniversary of the original entreaty by Emperor Kameyama-Jokō who had personally prayed at the Grand Shrine of Ise, asking for intervention on Japan's behalf by the great Shinto deity, Amaterasu ("who is said to shine in heaven"). Emperor Kameyama-Jokō is historically recorded as having made this entreaty on 15 August 1281, 664 years before the same day in 1945. This prayer had called in the Divine Winds known as the Kamikaze which destroyed the Khan's massive fleet of 3,900 ships and 140,000 Mongol marines.

United States Secretary of State, James Francis Byrnes, however, had been forced to draft a response that included the following: "...the ultimate form of government of Japan shall...be established by the freely expressed will of the Japanese people."

The Byrnes Note was sent to Japan on the 11th of August in 1945. On that same day, General George Marshall ordered Groves not to ship any more fissionable materials to the Pacific.

In America, most people believed that peace had come. "Japan Offers to Surrender," bannered the *New York Times*. Another *Times* story was headlinged: "GI's in Pacific Go Wild With Joy 'Let 'Em Keep Emperor' They Say".

In Japan, however, the war went on. The Japanese offer of peace and the Allied reply were known only to high government officials on either side of the Pacific. Morning newspapers in Japan on August 11 carried a statement in the name of General Anami that openly addressed the Army: "The only thing for us to do is fight doggedly to the end...though it may mean chewing grass, eating dirt, and sleeping in the field."

On August 12th, one of the God-Emperor's Uncles, Prince Asaka, asked whether the war would be continued if the Kokutai could not be preserved. Hirohito simply replied "Of course." Note, however, that Japanese is what is considered a "pro-drop" language – if the meaning is clear without one of the words, then you generally leave the word out, it being considered more elegant to imply things rather than say them directly. This meant, of course, that Japan could preserve its government. Hirohito had supreme confidence in his own ability to prosecute the war.

On August 13, 1945, Lt. Gen. John Hull (Assistant Chief of Staff for Operations and Plans) informed Col. Seeman that U.S. Army Chief of Staff George Marshall would proceed to use atom bombs only in direct support of tactical operations. Col. Seeman was expecting seven uranium "Little Boys" from post-cease-fire Germany by the time of Operation Olympic, the plan to invade Kyushu (Japan's most southerly of the main islands), however, it would take until July of 1946 before as few as six more Axis "Little Boy" bombs could be rendered available. Once a secure beachhead could be established by U.S. soldiers and marines atop rivers of glowing radioactive glass, an assembly-line of death was to ensue on an unprecedented scale. By late 1946, the "Rain of Ruin" Truman promised was finally intended to have been unleashed on Japan at the rate of over one "Fat" American plutonium bomb per week, serving as ordnance support for on-the-ground "line-breaker" offensive(s) while American troops were to advance and exterminate the civilian population, even though they were fully expected to lose all of their own personnel to ever increasing levels of radiation exposure. In spite of American enthusiastic bravado to drop more bombs, their fervor was tempered by factual circumstances.

After the attack on Hiroshima, Dr. Yoshio Nishina, an internationally recognized forefather of atomic research, was immediately flown to ground zero. His geiger-counter readings were coded in radio transmissions to Tokyo where Dr. Sagane Ryōkichi underscored his confirmation of the American escalation of war to nuclear level(s) with calculations as follows: "All of the American refining facilities and power plants that could be devoted exclusively to yielding fissionable materials – if allowed to operate twenty-four hours a day for seven days a week for three years — might be able to produce two or three atomic bombs."

Also, by reason of their offshore submarine-launched aerial reconnaissance and radiation-sensitive balloon trains, Japan's physics community knew that the enemy had already wasted one plutonium bomb in "testing" at the Trinity site and concluded that, after Nagasaki, if the Americans were not already out of domestically produced atomic bombs, they had only one more left in their domestic nuclear arsenal. This was known as the so-called "Third Bomb". Thus, a fourth domestically manufactured American atomic bomb would not be ready until mid-September of 1945 and realistically, not until October. In the critical month or two ahead, only the third bomb remained. This fact gave Japan a four month strategic window of opportunity to deploy their own (entirely unanticipated) weapons of mass destruction against America's homeland.

These calculations did not effectively factor in America's expanding nuclear industry that was already being devoted to plutonium production nor the booster shot given slightly more than two months earlier by a ship arriving in New York Harbor with over three hundred troy ounces of captured German uranium (refined to nearly ten percent purity of U-235). Even so, the latest American ground-based plutonium trigger tests had misfired in lopsided fashion – instead of in accordance with their impossibly precise spherical implosion design – and they fired prematurely, meaning that, if provided with a real plutonium core and dropped from a B-29, the triggers would have disintegrated both the bomb and the plane, and while American bombardiers were eager to kill, they were less willing to die.

The latest American uranium bomb casing and tamper fared no better. It met with misfortune when it (minus its priceless uranium) was dropped by mistake near a Chicago runway and therefore required a replacement to be built from square one.

Although not nearly approaching what I have suggested, mainstream academics have begun to demystify the lies concerning the propaganda around the atomic bomb. In other words, they understand and convey from history that the atomic bombs were not the true cause of the Japanese ending the war. Instead, however, some of them say it was the threat of invasion from Russia. As Japan is only a short distance from Korea, it is true that the Soviets could have reached Japan if they invaded all the way down the Korean Peninsula. What is avoided by historians, however, is how Japan maneuvered circumstances so that Russia was completely prohibited from invading all of Korea, let alone Japan itself. It is a great secret that is hiding in plain sight and it has to do with Japan's own nuclear program. I will address that in a subsequent chapter, but what is important to realize is that neither the Emperor nor his empire were deterred by either the atom bomb nor the Russians. At this point in the war, Hirohito's gravest threat was from within his own Empire.

CHAPTER TWENTY-TWO

THE PALACE COUP

Although I have portrayed Emperor Hirohito as an extremely intelligent, calculating and cunning individual, he was also someone who was continually challenged, and one of his greatest challenges was from his own people. Although Japan's military resources were formidable and they had only just begun to fight, there was a major problem with his military. As more and more troops were continuing to be mobilized, there were more than a few factions that did not favor the Emperor and for at least two very understandable reasons. As mentioned earlier, Hirohito had deliberately allowed his aircraft carriers to be sacrificed in favor of invading the Aleutian Islands. Not only were the carriers sacrificed but so were human lives. While this tactic is understandable from a strategic point of view, it is easy to understand how the rank and file sailors and even officers might feel betrayed, even if they were bound to serve the Emperor for the sake of the Empire. There was also the matter of the pre-arranged assassination of Admiral Yamamoto, an extremely popular celebrity amongst the nation who was also revered by the military. These actions did not sit well with a significant number of key military personnel.

On the morning of August 14, the situation with a surrender or peaceful terms between Japan and the Allies was no clearer, and there were very tenuous situations in play that could have caused a harrowing shift in the course of the postwar world. On this very night, the final night of the proactive war, as Emperor Hirohito recorded a "message of surrender" (which was really simply a cessation of hostilities) to be announced to the Japanese people, a band of Japanese rebels, commanded by War Minister Anami's elite staff, burst into the Imperial Palace. They had plotted a massive coup d'etat that aimed to destroy the recordings of the "Imperial Rescript of Surrender" and issue false orders forged with the Emperor's seal commanding the widely dispersed Japanese military to continue the war. On the very same day, however, in what amounted to a remarkable twist of fate, the Americans' final bombing mission of World War II would unwittingly prevent this military coup that would have continued Japanese participation in the proactive war.

As U.S. bombing missions resumed on August 14th, 1,014 bombers hit the Japanese mainland in what amounted to being the largest raid in the Pacific Theatre. These bombing missions included releasing another blizzard of leaflets that swirled over Tokyo and other cities, and this time they contained news of the messages exchanged between Japan and the Allies. Marquis Koichi Kido, Hirohito's closest advisor, later recorded in his diary that seeing one "caused me to be stricken with consternation" over the possibility that some leaflets could "fall into the hands of the troops and enrage them", thus making a military coup d'état "inevitable".

A coup against the Emperor was, in fact, already underway. While the War Minister, General Anami, was not involved in the coup, his elite staff were. If Anami were to give his support to the plot against the Emperor, much of the Japanese army, at least a million soldiers in the home islands, would almost certainly rise against him and the cabinet with the claim that the Emperor had been duped by cowardly civilians. In other words, the Japanese military were passionate about continuing the cause of the war. If Anami resigned from the cabinet, it would fall and Japan would fight on.

At Kido's frantic urging, the Emperor issued an Imperial Command from his air raid shelter: "I desire the cabinet to prepare as soon as possible an Imperial Rescript announcing the termination of the war."

Hirohito knew that the mere publication of such a rescript—a proclamation of the gravest import—would not be enough. He decided to be a true "Voice of the Crane".* He would step before a microphone and read the rescript to his people, none of whom had ever heard him speak before.

It is an important historical anecdote that on the 14th of August, after reading his prepared script, the Emperor asked the sound engineer, "Was it all right?"

The engineer stammered, "There were no technical errors, but a few words were not entirely clear."

The Emperor then read the rescript again and expressed that he would do so a third time if necessary. Although each reading was only four and a half minutes long, the speech spanned two records. The technicians picked up the first set of records for the broadcast, but they kept all four, putting them in metal cases and then into khaki bags. The technicians, like everyone else in the palace, had heard rumors of a coup and decided to stay there that night rather than attempt a return to the NHK national broadcasting studio, fearful that the army mutineers would attempt to steal and destroy the recordings. A chamberlain placed the records in a safe in a small office used only by a member of the Empress's retinue, a room normally off-limits to men. He then hid the safe within a pile of papers.

In the midst of all this and an "end-of-war" celebration on Guam, Air Force radio operator Jim Smith and his fellow crewmen received urgent orders for a bombing mission over Japan's sole remaining oil refinery north of Tokyo. The 315 Bombardment Wing was flying 3,800 miles to destroy

---

*The "Voice of the Crane" refers to the fact that a crane is often heard when it is not seen, often in the sky amongst the clouds.

the Nippon Oil Company refinery. As a stream of American B-29 bombers approached Tokyo, Japanese air defenses, fearing the approaching planes, signaled the threat of a third atomic bomb and ordered a total blackout in Tokyo and the Imperial Palace, completely disrupting the rebels' plans.

In the early hours of August 15, at Imperial Guards headquarters, the convergent tension expressed itself in the murder of both Lieutenant General Takeshi Mori, commander of the Imperial Guards Division, and his brother-in-law, Lt. Col. Michinori Shirashi. Major Kenji Hatanaka, a fiery-eyed zealot, burst into the office of General Mori with Army Air Force Captain Shigetaro Uehara and fatally shot Mori prior to butchering his body with a katana (samurai sword). Captain Uehara personally beheaded Lt. Col. Shirashi.

Two young officers looked into the murdered general's office as a third wiped Mori's blood from his sword. The hacked and bleeding bodies lay on the floor beneath him with blood still spurting from their wounds. The Lieutenant Colonel had his head sliced off, and the office walls were liberally splashed with red. The floor was both dark and slippery.

The murderer Hatanaka then affixed Mori's private seal to a false order directing the Imperial Guards to occupy the palace and its grounds and to sever communications with the palace except through Division Headquarters, occupy NHK (the radio station), and prohibit all broadcasts. This faction of the military was intent on stopping Hirohito's message of peace.

Meanwhile, Major Hidemasa Koga, a staff officer with the Imperial Guards, was trying to recruit other officers into the plot against the Emperor. At the palace, soldiers supporting the coup, with bayonets affixed to their rifles, rounded up the radio technicians and imprisoned them in a barracks. Wearing white bands across their chests to distinguish themselves from guards loyal to the Emperor, they stormed the palace and began cutting telephone wires.

Koga, hoping to find and destroy what he thought was the single record of the Emperor's message, ordered a radio technician to find it. The technician, unfamiliar with the palace, led several soldiers into the labyrinth. Soldiers roamed palace buildings, kicking in doors and flinging contents of chests onto the polished floors. The Emperor remained in his quarters and watched through a slit in the steel shutters protecting his windows.

Meanwhile, Lieutenant Colonel Takeshita tried to bring Anami into the plot. Anami declined, having already planned his own suicide to escape participation in the plot. With Takeshita in the room, Anami knelt on a mat, drove a dagger into his stomach, and drew it across his waist. Bleeding profusely, he then removed the knife and thrust it into his neck. Takeshita then pushed the knife deeper until Anami finally died. Meanwhile, Marquis Kido, Hirohito's personal advisor, was hidden from the plotters.

Rebellious soldiers swarmed into the NHK building, locked employees in a studio, and demanded assistance so that they could go on the air and urge the nation to fight on. Shortly before 5 a.m., on August 15, Hatanaka walked into Studio 2, put a pistol to the head of Morio Tateno, an announcer, and said he was taking over the 5 o'clock news show.

Tateno refused to let him near the microphone. Hatanaka, who had just killed an army general, cocked his pistol, but impressed by Tateno's courage, lowered the gun. Meanwhile, an engineer had disconnected the building from the broadcasting tower. If Hatanaka had spoken into the microphone, his words would have gone nowhere.

At the Imperial Palace, the Emperor was awake and listening as his court chamberlains debated how best to safeguard his majesty's life if it came to that. It was decided to bar all the doors and fasten all the Gobunko (Gobunko refers to the Imperial Library) shutters, but the latter had become so rusted over many years of nonuse that the chamberlains were unable to shut them. Finally, they ordered some tough young Imperial Guards to bar all the windows, and one chamberlain mused how ironic it was that this was being done against Japanese — and elite Imperial Guards at that — and not enemy soldiers.

Later, Premier Suzuki's residence was machine-gunned and Kido's villa was also attacked while Chamberlain Yoshihiro Tokugawa faced down his dispatched killers. According to one account, Tokugawa was assaulted. "A fist struck him full in the face, and he fell to the floor, his eyeglasses crushed beneath him...."*

At that point, the Imperial Library itself was surrounded by rebellious troops: "I still can't believe it!" asserted Fujita, who opened one of the iron shutters to peep out, seeing knots of Imperial Guards occupying key posts between the Fukiage** Gate and the Gobunko. "Machine guns were being placed with their muzzles pointing squarely at the house of the Emperor!" noted another account.

It took most of the night for troops loyal to the Emperor to round up the rebels. At dawn, they finally removed the mutineers from the palace grounds. Prime Minister Hideki Tojo's son-in-law, Major Hidamasa Koga, sliced open his own stomach in the sign of a cross, it was recalled later.

Now judging it safe to leave, the NHK engineers brought the Emperor's records to the radio station in separate cars using different routes. They hid one set in an underground studio and prepared to play the other. At 7:21 AM, Tateno went on the air, and without

---

*In 1960, a full 15 years after the events of August 1945, an Army sergeant sought out Chamberlain Tokugawa to offer an apology for having hit him, bringing as a peace offering an important family possession, a tea kettle made from a bronze mirror.

**Fukiage refers to the Fukiage Palace, the main residence of the Emperor.

recounting the adventures of the night before, announced, "His Majesty the Emperor has issued a rescript. It will be broadcast at noon today. Let us all respectfully listen to the voice of the Emperor. Power will be specially transmitted to those districts where it is not usually available during daylight hours. Receivers should be prepared and ready at all railroad stations, postal departments, and offices, both government and private."

At 11 AM, Hirohito was driven the short distance across the palace grounds from his living quarters to the blacked-out building of the Household Ministry which ran the affairs of the Imperial Family. In the audience hall on the second floor, the NHK technicians bowed to the Emperor.

An officer stated, "The Emperor's broadcast will be on the air very soon."

If the rebellion had succeeded, the military would have proceeded with large-scale Kamikaze attacks on Allied forces, costing huge casualties and just possibly provoking the Americans to drop a third atomic bomb on Japan over Tokyo and continue to drop more bombs as Japanese resistance stiffened. Additionally, this final bombing mission of the Americans is an insightful piece of a historical contingency investigation that explores how two seemingly unrelated events could have profoundly changed the course of modern history.

A few hours after the Americans had completed their bombing mission over Tokyo and the Nippon Oil refinery, the Emperor announced the end of the war over Japan's airwaves. The coup having been overcome, the Emperor's rescript would be heard, first by the Japanese and then, bogusly translated, throughout the entire world.

THE ROSWELL DECEPTION

CHAPTER TWENTY-THREE

THE NON-SURRENDER

At noon, on August 15, 1945, throughout all Japan, as the Emperor's voice was heard, people sobbed. Kazuo Kawai, editor of the *Nippon Times* wrote, "It was a sudden mass hysteria on a national scale."

The Emperor, however, spoke in a semi-archaic courtly or classical version of Japanese which was privileged to the nobility and was not readily understandable to most Japanese people. It was understood even less by the Allies, although they did not hesitate to spin it to their advantage for vital public relations purposes.

Most Japanese in 1945 would have found the Emperor's announcement very difficult to understand in the first place. Even those who were comfortable with classical Japanese were "much more used to reading language like that than hearing it spoken". The low quality of the recording did not help either. Historically, there was a follow-up broadcast which explained the announcement in plain-everyday-Japanese, but that is impossible to track down as the American military requested every copy for destruction so that it could never be coherently translated into plain American English.

All of this means that it is highly problematic to ascertain by this chronometric point-in-time how ordinary Japanese would have even "understood" the speech. My own mother, however, being highly educated and of the nobility herself, did understand the speech and it was most definitely not a speech of surrender. While the plain language explanation is not likely to see the light of day, what we do have is the officially translated version of the Japanese Imperial Rescript as it appeared on page #1 of the Nippon Times, on the 15th of August in 1945.

"The war situation has developed not necessarily to Japan's advantage, while the general trends of the world have turned against her interest. Moreover, the enemy has begun to employ a new and most cruel bomb. We have resolved to pave the way for a grand peace for all the generations to come by enduring the unendurable and suffering what is insufferable."

Even in the American version, Hirohito never used the words "defeat" or "surrender". Further, the words "surrender", "capitulate", "failure" and "defeat" are entirely absent.

The only reason the speech was read at all, however, was that President Truman had quite literally met the ultimatum of the Emperor by agreeing to a cease-fire in response to the super-dirigibles that had landed at Tonopah Army Air Field in Nevada, now known as Area 51. Deliberately surrendered into the hands of the U.S. Army, they were fully loaded with biocidal weapons that were entirely capable of destroying the entire human race, let alone the population of the United States of America. This understandably frightened the living hell out of the American high command as well as Truman himself. The President had no choice but to sue the Emperor for peace, all of which was done on back channels, the paper trail of which I was employed to destroy as an expert in documents destruction. Hirohito, who had been requesting an acceptable peace for years, had now set a final ultimatum and that was for them to acquiesce by the date of August 15th as this was the anniversary of the original entreaty by Emperor Kameyama-Jokō who, as has been described previously, had personally prayed at the Grand Shrine of Ise, asking for intervention on Japan's behalf by the great Shinto deity, Amaterasu ("he who is said to shine in heaven"). Emperor Kameyama-Jokō is historically recorded as having made that entreaty on 15 August 1281, 664 years before the same day in 1945. This prayer had called in the Divine Winds known as the Kamikaze which destroyed the Khan's massive fleet of 3,900 ships and 140,000 Mongol marines. This was an invasion fleet equal in size to the number of ships that transported and supported the Allied invasion fleet at Normandy.

Since the Emperor's speech was in Court Japanese (obscure even in 1945), exactly how to interpret it is not clear. Some modern commentators accept the "understatement" reading, per usage of 「必ずしも ("Kanarazu shimo")」, which some people claim to be the key word corresponding to "not necessarily"; but the cited passage of "things have definitely not gone well for us (which is signified to mean 'not necessarily' in the Americanist version)" has been acknowledged as grossly misleading over these many years, as Japanese have always known it to mean "despite everyone's efforts (which is implied) things will not necessarily improve for us."

Whether or not you believe in the divinity of the Japanese Emperor(s), a prospect which Hirohito himself wanted his people to disengage from, the coincidence of it all cannot and should not be dismissed. It is also true that the circumstances surrounding the palace coup required a lot of different factors to line up for the Emperor to succeed in the way that he did.

After the Emperor's broadcast, Major Kenji Hatanaka ended his mutiny standing outside the palace gates, trying to hand out leaflets that called on civilians to "join with us to fight for the preservation of our country and the elimination of the traitors around the Emperor." No one took the leaflets and Hatanaka shot himself in the head.

It is additionally noteworthy that the Imperial Rescript to the soldiers and sailors and officers issued on August 17th, 1945 makes no reference whatsoever to the atomic bomb(s) as these American terrorist attacks on civilians had no effect on military assets in the field.

THE ROSWELL DECEPTION

CHAPTER TWENTY-FOUR

The Aftermath and the Myth

Before we can assess the key facts surrounding the aftermath of what most perceive as the end of the war, I must first state and/or reiterate the major aspects of the myth that has been convincingly embedded into the historical narrative by the Office of War Information.

The Myth

Japan had no choice but to unconditionally surrender to the Allies by reason of the absolute devastation their population had suffered as the result of two atomic bombs being dropped upon Hiroshima and Nagasaki. Deeply apologetic for the crimes of his country, Emperor Hirohito put himself at the complete mercy of General MacArthur who determined that the best way to rebuild Japan and restore their economy would be to engender the good will of the Japanese people by retaining the Emperor, the symbol of the Empire who was deeply respected by the population.

A formal and official ceremony of surrender was signed by the Japanese aboard the battleship *Missouri* on September 2, 1945. Celebrated as VJ Day, this officially ended the war. (VJ Day stands for "Victory Japan Day" which is also sometimes designated as August 15, 1945, the day of the Emperor's transmission to the Japanese people that was actually a pronouncement of peace.)

Further, MacArthur guided the Japanese to turn their back on their old ways and helped them establish a new government based upon democracy where everyone could have a vote. As a result of MacArthur's leadership and the goodwill of the United States, Japan recovered to enjoy a thriving economy, thus becoming a staunch ally of the United States in the fight against communism, eventually leading to their inclusion in the United Nations.

These are the main points of the mythological narrative that most people subscribe to by reason of having it repeatedly fed to them by various forms of media, including movies. While this has been artfully orchestrated by the Office of War Information and its successors, nothing could be further from the truth. The fact that it is tacitly accepted as the truth is a testament to the effectiveness and pervasiveness of the effort to control people's minds and opinions.

The Truth

The most obvious and easy way to disprove the factual error of this myth is the widely believed idea that the Japanese unconditionally surrendered on September 2, 1945 aboard the *U.S.S. Missouri*. There was indeed a ceremony conducted aboard the *Missouri*, but as you can see below, it was signed in the presence of a rather unusual distortion of the American flag which is, in fact, an illegal flag that is deliberately hung backwards with all of the stars positioned upside down so as to emulate satanic pentagrams. There is a very bizarre back story to that particular flag which must be addressed in order to understand the entire context of what is known as the "unconditional surrender" aboard the *Missouri*.

This illegal flag was actually the same flag that Commodore Perry had flown upon his forcible entry into Japan. Aside from the stars being upside down and the bars being backwards, it is white on the other side. The reason Perry carried an illegal flag was that he was under orders to hoist this illegal flag in the event of any armed conflict that his fleet might encounter with any European naval powers. By hoisting this flag, the United States presumed that it could plausibly deny that he was acting in their interests with the intent to deflect responsibility away from the constitutional republic of the United States.

ABOVE IS A CLOSE-UP OF THE ACTUAL
ILLEGAL FLAG ABOARD THE *U.S.S. MISSOURI*

THE AFTERMATH AND THE MYTH

MACARTHUR IS IN FRONT OF THE FLAG WITH THE FRENCH, BRITISH, RUSSIANS, DUTCH, AUSTRALIANS, AND NEW ZEALANDERS, BUT YOU DO NOT SEE A SINGLE CHINESE BECAUSE THEY WERE NOT PART OF THAT CEREMONY AS THEY HAD REALIGNED THEIR STATUS AS AN AXIS NATION.

The reason that the flag was white on one side was that, if need be, Commodore Perry was authorized to surrender his fleet by reversing the colors and presenting a sheet of white, thus consigning himself and his men to hanging in the event of overwhelming odds. In such an instance, Perry and his fleet would be repudiated as a rogue fleet under the command of a renegade admiral. At that time, the United States was an inferior naval power and still smarting from the Capitol having been sacked and the White House itself having been burned to the ground during the War of 1812. Seeking to avoid inciting their rivals on the high seas, they wanted to be able to deny their colonial encroachment upon Japan as well as absolve the United States of any involvement in his actions and thereby defer a concomitant state-of-war with any great European power or alliance thereof.

Another peculiar and undeniable aspect of the history of this flag was not only that it was used for the *Missouri* ceremony but the circumstances under which it was procured. It most certainly was couriered with overwhelming urgency. In fact, it was a rushed delivery that spanned 9,500 miles in 120 hours by a special courier whose orders required him to keep the wooden box containing the false flag in his sight at all times. This meant sleeping with it, eating with it, and even taking it with him to the lavatory. Upon arrival on the *U.S.S. Missouri* and delivering the false flag, the special courier slept for forty-eight hours straight and missed the eighteen minute event.

We have to ask: why was it so important that this flag was displayed on ceremony?

As the ceremony aboard the *Missouri* was not legitimate, the flag that was displayed on ceremony was one that could not be legally recognized. This is because the U.S. Federal Government's claim(s) to its own citizenry were false. The ceremony was only to SYMBOLIZE perceptions towards proactive cessation of hostilities. It was neither a cease-fire nor a surrender although it would be misrepresented as both in America. Although official cessation of hostilities or even cease-fire would not be officially recognized until December 31st, 1946 as per Title 38 of The United States Code, there have been tremendous efforts to spin the truth so that it appears otherwise, i.e., as an "Instrument of Surrender". The following document is the actual and formal declaration of cessation of hostilities as declared by President Truman on the last day of 1946.

Proclamation 2714: Cessation of Hostilities of World War II

December 31, 1946

By the President of the United States of America a Proclamation:

With God's help this nation and our allies, through sacrifice and devotion, courage and perseverance, wrung final and unconditional surrender from our enemies. Thereafter, we, together with the other United Nations, set about building a world in which justice shall replace force. With spirit, through faith, with a determination that there shall be no more wars of aggression calculated to enslave the peoples of the world and destroy their civilization, and with the guidance of Almighty Providence great gains have been made in translating military victory into permanent peace. Although a state of war still exists, it is at this time possible to declare, and I find it to be in the public interest to declare, that hostilities have terminated.

Now, THEREFORE, I, HARRY S. TRUMAN, President of the United States of America, do hereby proclaim the cessation of hostilities of World War II, effective twelve o'clock noon, December 31, 1946.

IN WITNESS WHEREOF, I have hereunto set my hand and caused the seal of the United States of America to be affixed.

DONE at the City of Washington this 31st day of December in the year of our Lord nineteen hundred and forty-six, and of the Independence of the United States of America the one hundred and seventy-first. [SEAL]
HARRY S. TRUMAN

By the President:
JAMES F. BYRNES
The Secretary of State

Fully aware that the ceremony aboard the *Missouri* was neither a cessation of hostilities nor a surrender, it was the Japanese who demanded that this illegal flag be displayed as a backdrop to the ceremony because, when Commodore Perry had originally invaded Japan in 1852, he lied to them that the upside down banner actually represented the United States. While they fully allowed the Americans to propagate their false PR, it was the intention of the Japanese to "close the circle" on this whole episode by conducting the ceremony in Tokyo Bay, the very same location where Perry had invaded 93 years prior, under the identical flag that Perry had carried with him.[*]

Also playing off of the fact that MacArthur promoted himself as a relative of Commodore Perry, the entrance of the American fleet and the *U.S.S. Missouri* into Tokyo Bay, with the same flag, was meant to signify the opening of Japanese markets to the world. In other words, the Japanese had been forced under threat of violence to participate in international commerce in 1852; and now, after having been excluded from such under the threat of colonization, the "rest of the world" (the Allied countries) were now opening up their markets to a liberated Japan.

There is additional symbolic representation in the closing of this circle, and that has to do with President Harry Truman's home state being Missouri, after which the battleship of the same name was christened. There is also undeniable incidental occult symbolism. The number of

---

[*]It is a matter of historical record that, after I came out as a public informant circa 2011 and pointed out the exact circumstances surrounding the illegal flag used by Commodore Perry and its presence aboard the *U.S.S. Missouri*, the military completely changed its tracks with regard to using bastardized versions of the American flag, issuing patches and the like where the American flag would be backwards or appear in an irregular fashion. Previous to this, any desecration of the flag was expressly forbidden and considered unpatriotic or subversive. In order to maintain and perpetuate a state of ignorance, they were obviously on high alert for damage control purposes and ready to respond to any information I might release.

inverted stars on the flag is 31, reflecting the number of states in the union at the time of Commodore Perry, instead of 48, the latter being the number of states in the union circa 1945. There is also the fact that the "closed circle", from 1852-1945, represents 93 years, 93 being the number for the Greek THELEMA or (FREE) WILL, an identification that was forever spouted by occultist Aleister Crowley.* It is also of numerological significance that the *Treaty of San Francisco* went into effect in 1952, exactly one hundred years after Perry's incursion.

Although the Emperor was well aware of the ceremony and did not order the instrument to be signed, he allowed it to take place but did not dignify it in any way by either attending it or even acknowledging it. Besides making the above demands, he was playing off of the ego of his adversaries, and it was a cunning and strategic maneuver that allowed his enemies to save face by staging a public relations coup for their own people. The Americans even sold tickets to the event which, as was previously alluded to, lasted only eighteen minutes.

As for the so-called "Instrument of Surrender" itself, it is a point of fact that the U.S. Government itself does not recognize it save for public relations purposes, it having been abrogated at time of signing as Russia and Japan were still in and yet remain to this day in a declared state of war, thus obviating the entire document as moot. You stand as Allies; you fall as Allies.

At the time of the so-called "surrender", the Japanese had just defeated the Soviets in the Battle of Shimushu. At that point in the war, the Soviets, who had suffered greatly in the European Theater with most of their young men now dead or otherwise deployed, were able to spare but an ill-equipped and completely ridiculous rag-tag collection of old World War I cavalry veterans waving sabres on horseback. In their effort to reach Hokkaido, they were using paddles and life preservers, their landing craft having been blown out of the water by Japanese amphibious Type 2 Ka-Mi tanks.

Remaining openly hostile to the Japanese for years, the Russians never even signed the *Treaty of San Francisco*, the document which actually ended the war, signed on September 8, 1951 and purposely designed to go into effect on April 28, the Emperor's birthday. This is now a national holiday in Japan, and Hirohito, the Showa Emperor, is the only Emperor in Japanese history to be honored with a national holiday every year. This indeed represents a national day of celebration by the

---

*Besides being an inversion of the number *13*, the number *31* is also a key number in Aleister Crowley's seminal work, *The Book of the Law*, wherein it is referred to as *Libre 31*, its assigned number. Separately to this, the number *31* is also considered a key to understanding the book in terms of Christian Cabala.

Japanese in honor of someone recognized as a divinity who saved their Empire from destruction by foreign invaders.

There have also been tremendous efforts by the Office of War Information and its successors to spin all of these facts so that they can be misconstrued as to align with the overall myth that has been perpetrated upon the world and the United States in particular. It is, however, a historically observable fact that the document which actually ended the war was the *Treaty of San Francisco* that was signed six years after the so-called "Instrument of Surrender".

Referring back to *Proclamation 2714* and the cessation of hostilities, this proclamation was on behalf of the United States and showed demonstrable good will and endorsement of the United Nations, thus committing to "no more wars of aggression calculated to enslave the peoples of the world and destroy their civilization". Of even further significance is that Truman states unequivocally that "a state of war still exists".

All of these facts are evident five years after this proclamation during the very first coast-to-coast transcontinental televised broadcast in American history which took place on September 4th, 1951. The broadcast was to signify and address peaceful relations between the United States and the greater Japanese Empire after six long years of peace talks and negotiations. You can only find the first minute or two of the original introduction by Truman on youTube, but even though it was the very first coast-to-coast and first transatlantic broadcast in human history, this seminal transmission cannot be found anywhere on Google. It has been accordingly scrubbed because Truman's speech announces to the world that the United States has succeeded in the reconstruction and rebuilding of Japanese industry and would open up their markets to the Japanese Empire as a specifically privileged trading partner. As the Americans had placed an embargo on supplying oil to Japan and had effectively imposed sanctions on their trading capabilities, Truman's capitulation was even more than what the Japanese had originally sought prior to the war. The Truman broadcast was scrubbed and censored on the internet because it proves the war of hostilities between the United States and Japan ended on September 8th, 1951.

While you can find a transcript of Truman's September 4th speech on the internet and the rhetoric used by Truman favors the public relations position of the United States, it demonstrated that the war did not end in 1945 as the President also addresses the success of the Marshall plan on the reconstruction of Japan. Keep in mind that what the Americans reference as "Occupation", the Japanese refer to as "Reconstruction".

This signing and declaration of peace reveals the withdrawal of American presence in Japan and exposes the fact Japan had won the war without any unconditional surrender nor any legal consequences, and in fact, it also demonstrates how the United States paid billions of dollars in reparations to the Japanese Empire, ending the war in 1951 with the peace treaty going into effect on April 28th, 1952, making Japan the wealthiest nation on Earth, all of this verifiable during this televised transmission.

As far as war reparations, the Americans were both obligated and forced to clean the atomic mess they had made, not the Japanese. It was American marines and army divisions who were sent in to clean up the atomic blast sites at Hiroshima and Nagasaki, the first large group of American Soldiers arriving in Nagasaki on September 23rd of 1945 and in Hiroshima two weeks later. This was a prompt response following cease-fire per demands of the Japanese government in order to prevent overexposure to any additional Japanese personnel or civilians. These soldiers were part of a force of 240,000 that occupied the islands of Honshu and Kyushu. Marines from the 2nd Division "took" Nagasaki while the U.S. Army's 24th and 41st divisions "seized" Hiroshima.

Without safety-gear and without anyone being urged to take precautions, the U.S. G.I.'s were ordered to bivouac close to Ground Zero and sleep on the very earth that had been blasted and irradiated. This was all at the behest of their own government as the latter wanted to specifically measure rates at which their hair fell out and how long it took for their bodies to become covered in sores. The Government considered it advantageous to use Americans as mass-radiation-exposure subjects because their fluency in English and lack of cultural barriers enabled U.S. Federal scientists to immediately comprehend their descriptions of symptoms as they deteriorated.

When an American marine named Sam Scione returned to the United States a year later, his body was covered in sores and all of his hair fell out. Suffering a string of ailments, he was never even awarded service related disability status, being but one among tens of thousands of such cases. Those soldiers and marines died by the thousands of radiation cancer. None of this, of course, is the position of a victorious power at all, but do Americans know of this? No, they do not, but if you look up "Atomic Veterans", you will discover that everything I am saying is true, and this is how the war went down.[*]

---

[*]Please Reference: "Radiation Dose Reconstruction; U.S. Occupation Forces in Hiroshima and Nagasaki, Japan, 1945—1946 (DNA 5512F)" and "DTRA Fact Sheets: Hiroshima and Nagasaki Occupation Forces."

CHAPTER TWENTY-FIVE

JAPANESE TECHNOLOGY

The entire history of the final years of the war (1945-1951) consists of many complex branches and facets of circumstances, all of which are obscured, not only by the sheer vastness of the history itself but by the efforts of the Office of War Information to focus on the major mythological war story as previously given. While entire books could be written on this subject, an overview of certain key Japanese technological capabilities is in order so that you have at least a grasp of the enormous capacity of the Japanese war machine that the Americans were facing, all in addition to the biological weapons of mass extermination.

The History Channel once featured an exclusive documentary, *Secret Japanese Aircraft of World War II*, which demonstrates that Japanese technology was far superior to anything the Allies possessed. The documentary is qualified only by its purposeful disinformation as it feeds the old agenda of the Office of War Information, suggesting all this technology came too little too late and that the Japanese were totally defeated before it could make a difference. The documentary does corroborate, however, how advanced the Japanese were compared to the Americans.

A prime example of Japan's superior technology was viewed by the whole world in 1937 when a series of headline-grabbing flights proved the genius of Japan's aeronautical acumen. What promulgated so much of the attention was that royal houses from across the globe were traveling to London for the coronation of Britain's King George VI and his wife Elizabeth. From Japan, Emperor Hirohito dispatched, by sea, the Crown Prince and Princess as his official envoys. In the background of this was an accomplishment in Japanese aeronautical engineering that would have truly amazed any Westerners who had heretofore considered the Far East to be technologically backward.

To cover this monumental wedding, a Japanese newspaper sponsored the flight of a pair of Japanese airmen, Masaaki Iinuma and Kenji Tsukagoshi, to make the ninety-six hundred mile journey in a state-of-the-art airplane, the Mitsubishi Ki-15.

In an aeronautical feat that earned them the acclaim of being known as the "Japanese Lindberghs", Iinuma and Tsukagoshi made the trip from Tokyo to London's Croydon Airport in only seventy-two hours, the three-day flight being between the 6th and 9th of April. Their route crossed broad tracts of jungle and open ocean, flying five and six-hour legs against the prevailing winds in tropical heat and humidity on seat cushions that left them shifting in pain.

Leaving Japan at 2 a.m., they landed in Taipei, taking off again at 10:20 a.m. for the westbound jaunt to Hanoi. After an hour on the ground, they were again airborne, finally ending their day in Vientiane, Laos. With four only hours sleep, they launched again, crossing the Bay of Bengal and northern India to Karachi where they lodged with Japanese expatriates. The next day, they crossed Iran and reached Athens with brief stops in Basra and Baghdad. Comparatively short hops took them to Rome and then to Le Bourget airfield in Paris. When they finally reached their destination, landing at Croydon on the outskirts of London, they had covered the ninety-six hundred mile route in a flying time of just fifty-one hours and nineteen minutes – a new world record.

Having arrived a month early, Iinuma and Tsukagoshi used the time between their arrival and the King's coronation to make a goodwill tour of Europe, visiting Brussels, Berlin, Paris, and Rome before returning to England. Welcomed with banquets, receptions, and photo ops with VIPs, they were hailed as international stars. Leaving for home on the fourteenth day of that month, the pilots delivered photos and accounts of the London coronation scene to their sponsor in Tokyo. In the process, they became national heroes.

The flight of the Mitsubishi Ki-15, powered by a reliable nine-cylinder engine and nicknamed "Kamikaze" (despite the name, it was not a suicide plane, the term *Kamikaze* meaning "Divine Wind"), was the first of an impressive series of Japanese distance and endurance flights during the buildup to World War II in the Pacific. Even with fixed landing gear, the Mitsubishi managed a top speed of three hundred mph and a range of thirteen hundred nautical miles. The U.S. Navy's frontline fighter at the time was the chunky Grumman F3F biplane, and Britain's Royal Air Force had just started flying the Fairey Battle which was half-a-hundred mph slower than the Mitsubishi despite having three hundred more horsepower and retractable landing gear.

Despite these heralded achievements, they did not even register as relevant to Western Intelligence observers at the time and are still completely unknown to Westerners today, and this is a very big deal. Not only is the aeronautical technology of Japan overlooked and denied, it was completely revolutionary for its time. After the *Kamikaze* was forcibly ditched in 1939 in the ocean off Taiwan, it was laboriously retrieved due to the revolutionary nature of what had been accomplished with the engineering as it was a source of Japanese national pride. Put on display near Osaka until 1947, the U.S. Army destroyed it to erase any evidence that would betray unto their fellow Americans how much more technologically advanced the Japanese had always been – even before engaging the West in open hostilities.

What is even more critical concerning the advancement of Japanese aeronautics concerns a modification they did of Germany's Junkers G38, a plane that was originally designed for carrying as many as a thousand people. The Germans had blended flying-wing technology of the period with a conventional fuselage and efficient tail assembly (referred to as empennage, a term deriving from the French *empenner*, "to feather an arrow", it being suggestive of the way the path of an arrow is stabilized by reason of the feathers on its tail). During the early part of its flying career, the Junkers model G38 was the largest land-based aeroplane anywhere in the world.

Under conditions of the strictest secrecy and with manufacturing license, the Japanese collaborated with Junkers engineers to develop a military-minded bomber-transport variant. Designated as the Ki-20, six of them were completed spanning from 1931 through to 1935 as the Army Type 92 Super Heavy Bomber. It is noteworthy that in the years prior to the war, the Japanese had produced three times the number of these revolutionary interwar craft than the Germans themselves did.

The Ki-20 became the largest aircraft to be operated by the Imperial Japanese Army Air Service branch. It was so big that crewmen were able to walk standing INSIDE the hollow singular wing of this aerocraft. As a measure of its size, the Ki-20 had a wing area double that of the Boeing B-29 Superfortress, first flown a decade later that would drop America's atomic bombs. Besides being decked out with machines guns positioned to fire at all angles, it was designed to carry up to 5,000 kg (11,025 pounds) of conventional bombs on external racks under the fuselage.

These ten-man superbombers served with the Imperial Japanese Army Air Force and the initial pair of them were assembled from Junkers-provided components by Mitsubishi with the first one flying "Kuro" (Kuro literally means "black", such a designation referring to the fact that it was being flown covertly) with German assistance in October of 1931 from Hamamatsu to Pyongyang (in Korea), the trip continuing for thirteen hours in the air while training. The first official flight would not be recorded until 1932. Whereas the first two Ki-20s were assembled from materials imported from Germany, the remaining four aircraft of the lot followed from 1933-1935 and relied evermore heavily on Mitsubishi-manufactured components, the last two planes consisting only of Japanese made parts. These planes became absolutely crucial to the Japanese as they began to roll out their atomic program.

Dr. Yoshia Nishina at his cyclotron

Although it is not really unknown to history, Japan's role as the premier researcher in the field of atomic warfare is both overlooked, under appreciated and even ignored. When Albert Einstein approached FDR to build an atomic bomb, he told him that the Third Reich and the Japanese were so far ahead in this endeavor that Roosevelt's science advisor, Enrico Fermi, told FDR it would take a hundred years to build an atomic bomb. As a result, the Americans lost the atomic race, something they will never tell you. American research into the atomic bomb began very late in the war whereas Japanese and German research began in the 1920s.

It was Yoshio Nishina who developed the atomic bomb. Internationally recognized, Nishina visited some European universities and institutions and did research with Niels Bohr. In 1929, he returned to Japan where he created the second cyclotron in the world, the first being in Germany, and established Nishina Laboratory at RIKEN in 1931 in order to study quantum mechanics, inviting Western scholars to Japan, including Heisenberg, Dirac and Bohr.

Japan's efforts with regard to their atomic program were boosted considerably as a result of their association with a powerful Jewish banker named Jacob Schiff who had bailed them out when they went broke fighting the Russo-Japanese War of 1904-05. Schiff backed the Japanese because the Russian Tsar was anti-Semitic, and Schiff's own family had been wiped out (in Russia) as a result. Able to take the war to a victorious conclusion and because of their national samurai sense of obligation, the Japanese demonstrated their appreciation three decades later by taking in over a million Jewish emigres from the Third Reich. This exodus was facilitated by Sir Victor Sassoon, a British nobleman from a wealthy international merchant family. His own family having

JAPANESE TECHNOLOGY

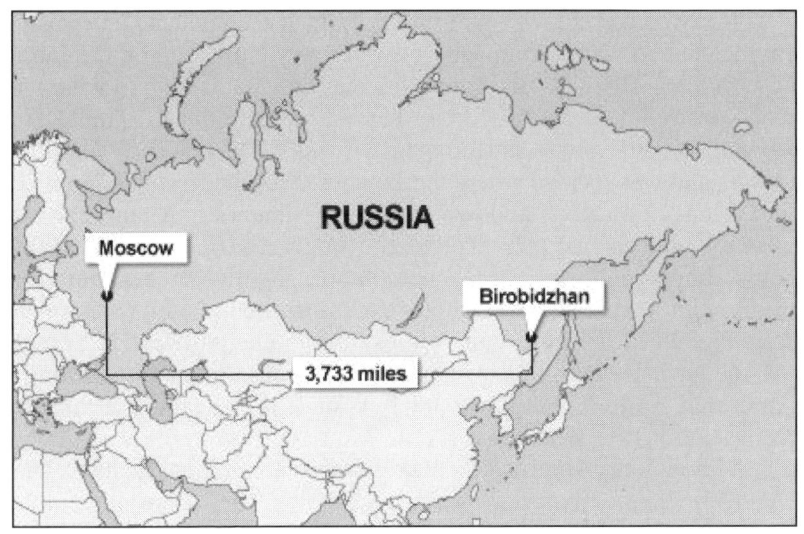

ON THE RIGHT IS BIROBIDZHAN, THE CAPITAL OF THE JEWISH OBLAST THAT WAS ONCE THE NATION OF YEVREY

migrated from Iraq to India, he eventually moved his empire to Shanghai to avoid taxation. Building an extensive real estate empire with high profile and luxurious hotels, he created an international financial district in Shanghai which had its own government and was respected by the Chinese because of all the wealth it facilitated. One of his jewels was the Shanghai Stock Exchange which was run almost exclusively by his fellow Sephardine Jews. As Hitler encouraged Jews to emigrate from the Third Reich, Shanghai remained open to them as a sanctuary port thanks to Sassoon's influence. Although the Japanese had invaded China, they respected the international boundary and Sassoon himself prior to declaring war on the United States and Great Britain.

In accordance with their sense of honor, the Japanese relocated over one million Jews to a little known town known as Birobidzhan, the administrative center of what was and is still a Jewish Autonomous Oblast.* Originally established by Emperor Hirohito in 1931 and later absorbed by the Soviet Union during their invasion of Manchuria, it served as a place for persecuted practicing Russian Jews, none of whom were welcome in the rest of the Soviet Union. The national language is Yiddish, and it was literally another precursor "Israel" in Asia, but it will not show up on any American maps.

---

*An oblast is a political subdivision of Russia. The term applies to either Imperial Russia, the U.S.S.R. or the Russian Federation. The original national name of this oblast was Yevrey.

Many of these Jews were highly trained technicians, physicists, engineers, and accountants who were happily employed by the Japanese in a massive industrial complex that they had set up in what we recognize today as North Korea. Known as the Konan Complex (in Korean, it is referred to as Hungnam), it was the largest such complex in Asia and was located where the Japanese had successfully dammed the Ch'ongch'on River and two other river tributaries. All three rivers were damned so that they generated enough electric power that they could almost run the entire Japanese nation. North Korea is saturated in uranium, and it was with this weapons grade uranium (the power sourcing out of the three dam gorges) and the help of Jewish Yevrey that the Japanese built scores of nuclear warheads. Every single atomic bomb that North Korea blows today is built in the Japanese nuclear facilities from the World War II era.

Although the Americans knew very well about these facilities and were desperate to bomb the Konan Complex, they were never able to pull it off because the Japanese islands were in the way. So many Americans were shot down trying to reach the Konan Complex by flying over Japan that no less than 5,000 bombardiers had to parachute onto the islands of Japan itself. Less than eighty of them made it out alive because they had killed so many Japanese civilians that the latter would literally tear them apart while still alive whenever they landed on the ground. The eighty that made it out alive were the lucky ones who were captured by the military or police for interrogation.

The fact of Japanese superiority in atomic research even revealed itself in the popular press as early as 1946 in the October 3rd edition of *The Atlanta Constitution*. While it does not have the full story and is only based upon the story of an American soldier/journalist, David Snell, having interviewed a refugee Japanese soldier, it corroborates a substantial number of points that I have to share based upon what I read in the classified files I was ordered to destroy. On the whole, it is a very accurate article.

Some people have been rather obsessed with trying to either compensate or dismiss the article by David Snell in *The Atlanta Constitution*, but it is not the only independent account of the Konan complex. One example is from the *Canberra Times* (see page 206 and 207) of 28 November 1950 which included the following article on the front page.

Although most of the Konan complex was underground, it was abandoned as part of a series of strategic maneuvers by the Japanese to use the Russians as a buffer with the Americans. At this point, the Japanese had their K-20 long range bombers fully equipped to deliver nuclear bombs if needed. At the same time, the Japanese had also

JAPANESE TECHNOLOGY

developed the Ki-77 and Ki-74, both capable of flying non-stop from Tokyo to New York.

In what appeared to be a defensive retreat to protect his homeland, Emperor Hirohito sought to bait the Russians by withdrawing the world's First Atomic Bomber Command from Manchukuo (Japanese Manchuria) for redeployment on the Japanese home islands. At the same time, American President Harry Truman had promised Stalin to provide the

THE ABOVE ARTICLE (SEE OPPOSITE PAGE FOR CLOSEUP) REVEALS A REPORT OF A JAPANESE ATOMIC PLANT IN KOREA

Soviet Union air support by-way-of reallocation of thirty-seven hundred B-29 bombers and other resources to the Pacific Theater, a commitment that was originally sincere but ultimately unfulfilled. All of this was enough to lure Stalin into taking the Emperor's bait and tossing his final assets into the Far East.

Launching Operation August Storm, its ultimate purpose being the invasion of Japan, the Russians invaded Manchuria as the Japanese

JAPANESE TECHNOLOGY

retreated to the south. The Russians followed them, their primary focus being to take the entire Korean Peninsula, it only being a very short distance from there to Japan itself. Taking a number of nuclear warheads from their arsenal and setting them off like atomic land mines, the Japanese literally disintegrated the Soviet advance. This eventually set the stage for the Korean War, a war that is still in force to this day, and actually created the demilitarized zone on the 38th parallel, the disputed dividing line between North and South Korea. Although no one talks about it, the remnants of the Konan complex and its legacy is what has enabled the North Koreans to have a nuclear program.

The demilitarized zone between what is now North and South Korea is the largest demilitarized zone in the world, an artificially man made area that is so large that it can be observed from the moon without aid. It is the only thing you can see from the moon except for the Great Wall of China. Industrial pollution, however, has inhibited this view. But, we have to ask: How was that demilitarized zone created? Nobody knows! Nobody talks about it. It was created by the Japanese when the Soviets were rolling down the Korean peninsula towards Japan itself. They created a demilitarized zone, a trench so large that it eventually became a nature preserve. There are several species of Siberian tiger, Southeast Asian tiger, and types of cranes that are important to many royal family lines in Asia, from the Ching to the Yomato Dynasties. All of these animals now thrive in this massively wide demilitarized zone created by dozens of nuclear warheads blown in the Korean Peninsula. Like Hiroshima and Nagasaki, they were short generation bursts of radiation that became safe within a period of a few months. Just like in Chernobyl, animals moved back in. With all the humans having departed, Chernobyl is now a huge game preserve for wolves, cats, mice and various other types of animals.

Even after being embarrassingly outmaneuvered in Korea, Stalin stupidly attempted an amphibious invasion of Japan's northern island of Hokkaido. This is where the Japanese naval tanks blew him out of the water. If you go to Hokkaido, you can find all these remnants of the busted amphibious ships, busted Soviet tanks, and also skeletons of horses and men that they were trying to transport as cavalry. Their boats were blown out of the water by Japanese naval tanks that were held in reserve, but Americans never speak of this. The Japanese never used banzai charges nor Kamikaze tactics. Consequently, Stalin was unable to invade Northern Japan like Truman was begging him to do.

The Japanese had accomplished what they wanted. Although the Russians were clearly defeated, they were a menacing presence in the Far East as far as their American allies were concerned. In a postwar world, the Americans would be wary of a communist presence, all of which would tip the balance of power towards Japan. In other words, Hirohito correctly calculated that the Americans felt far more threatened by the Russians than by his own empire. Although this was a very important card to play, it was far from the only one.

As for the land invasion, the Americans had a horrible time in Asia and were in a very perilous position if they were to press further. The Japanese islands are longer than the continental United States itself and have more land area than France or Germany. They are an enormous amount of land and everything the Americans had fought for up to that time had been a tiny speck of land. On these other coral reefs and atolls, Japanese and American soldiers were killing each other with bayonets and samurai swords on choke points where the ocean on either side was but fourteen yards apart. Suddenly, upon invading Japan, the Americans were at a loss because, for the first time, they were not only fighting on a very large territory but they were also outnumbered. Until that point in the war, as previously said, the U.S. had outnumbered the Japanese two to one but now the Japanese would easily outnumber them by a ratio of ten to one. Everything was changing at the end of the war. All the calculations were changing.

The original Allied plan was to kill all of the Japanese and split the Japanese islands with a joint administration between Nationalist China, the U.S., the U.K. and with the Soviet Union getting the northern half of the island so that they could finally have an ice free outlet to the Pacific, something they never realized. This plan, however, was nullified by Yoshio Nishina and his development of the atomic bomb.

It was also a fact that the Nagasaki bombing fiasco had proven conclusively that both the Japanese Imperial Family – and that twenty-six hundred year-old dynasty's industry – were more than adequately entrenched enough to survive America's "strategic" approach to nuclear warfare. Well over twenty-five hundred young ladies employed in Mitsubishi's underground-bunkered production facilities beneath the city of Nagasaki survived the atomic attack without even realizing there was any kind of disturbance overhead, thus continuing unabated full-scale assembly of Tachikawa Ki.74 "Patsy" long-range bombers capable of reaching New York City from Tokyo. It was thereby concluded (correctly) that any atomic attack on the Japanese national capital of Tokyo would leave the Imperial Bunker(s) equally unscathed.

There was also the potential to use both submarines and the super-dirigibles themselves for atomic warfare, the latter utilizing parasite aircraft.

As formidable as the atomic weapons were, an additional and rather astonishing testament to Japan's superior military position came to light

ILLUSTRATION OF PARASITE BOMBERS WITH ATOMIC WEAPONS

in the first decade of this 21st Century, around 2005-6, when a WW II-era Japanese Army fortress was rediscovered in Aomori Prefecture in northern Japan via an old access-tunnel. Inside was a fully-assembled super-battery of microwave-generation magnetrons equipped with magnification-direction apparatus which was, in fact, an operational death-ray. The Japanese government immediately disassembled this apparatus and confiscated the internal mechanisms as a potential safety-hazard. This was at least one of the heretofore undisclosed end products of the Imperial Japanese Army's high-frequency electric wave weapons project which, by 1945, involved 116 military technicians under the command of Major General Yukaba Hideki (also identified as Kusaba Sueki). Major General Kusaba had resumed oversight of microwave research at Noborito Laboratory around November 1944 after serving as the Noborito Labs Director responsible for supervising development and production of balloon bombs (Fu-Go) for firestorm generation within the continental United States. Major General Kusaba's personnel at Noborito included twenty technical officers, four civilian engineers, twelve part-time civilian consultants, and eighty technicians. It should be further noted that Yukawa Hideki won the Noble Prize for conducting research in physics on the electromagnetic pulse in 1949 while the Japanese and Americans were still at war.

Japan's advanced technology and the many branches of "big science" were a very important element with regard to how the war ended. In fact, Japan had its own "Area 51" that was dedicated to their advanced technology. It was known as "Monster Island".

CHAPTER TWENTY-SIX

MARCUS ISLAND

One of the most enigmatic and overlooked historical aspects of World War II concerns the hidden aspects of a small coral atoll in the Pacific, just under a thousand miles from Tokyo, known as Minami-tori-shima. Dubbed "Monster Island" due to on site development of unconventional weapons with maximum threat-projection potential, the Americans also refer to it as Marcus Island. Serving as a test bed for revolutionary prototype technologies, some of which you have read about herein, Minami-tori-shima has a reputation for being Japan's "Area 51" when it comes to advanced technology.

As Japan's most formidable vanguard of defense and offense when it came to weapons of war, this small atoll was attacked virulently and repeatedly by the Americans, the first coming on March 4, 1942 by the *U.S.S. Enterprise* in the Marcus Island Raid. It was also a primary target during Jimmy Doolittle's raid in June of the same year.

No less than four U.S. Submarine-Service Special Reconnaissance Missions dispatching UDT (Underwater Demolitions Team) runs were conducted against its facilities. The first two SCUBA Reconnaissance Raids, which were very risky for the frogmen, were in June of 1943 (*U.S.S. Searaven*, Commanding Officer H. Casedy) and August of 1943 (*U.S.S. Sunfish*, Commanding Officer R.W. Peterson). These confirmed that the advanced Japanese Weapons Programs at Marcus Island were deemed "war-breakers" in terms of their expected impact on the United States war effort.

Not only were the weapons sophisticated and beyond anything the Americans had in their own arsenal, the manufacturing fortress on Monster Island was impregnable. As the entire island was surrounded with a sizable coral reef, it was highly resistant to enemy landings of any sort. There were natural huge cavities in the coral which the Japanese further excavated, and this included a canal carved-out of the solid coral to permit uninterrupted submarine resupply throughout the Second World War.

After the high command had received the reports of the Special Reconnaissance Operations, they planned for an all out assault on Marcus Island. On August 31st of 1943, Rear Admiral Charles Powall commanded Task Force 15, a huge armada consisting of three carriers: the *U.S.S. Yorktown*, the *U.S.S. Enterprise*, and the *U.S.S. Essex*;

plus the battleship *U.S.S. Indiana*, two cruisers, at least four flotillas of destroyers, and even a fleet oiler to enable maintenance of an extended siege. It was the largest Aircraft Carrier Task Force ever deployed on a single mission in the Pacific throughout the Second World War, the intent of it being to obliterate the Monster Island Garrison (and its advanced-technology weapons development capacities).

Task Force 15 was followed up the next year when, on May 23rd, 1944, another impressive U.S. Navy Aircraft Carrier Task Force deployed a forty-eight hour 'round-the-clock attack on Monster Island with literally hundreds of planes.

Four months later, on September 7th, U.S. land-based liberators hit Monster Island for the first time in the Greater Pacific War. Marcus Island, however, was the home of the Japanese 16th Tank Regiment (during World War II, this was comprised of Type-95 Hai-Go's). Two further U.S. submarine-launched SCUBA-insertions of frogmen revealed that weapons development continued apace on Monster Island with no impediment(s) noted in the operational capabilities of the Imperial Japanese 16th Armored Regiment's abilities to repel invasion. This island was clearly invulnerable. Monster Island was never successfully garrison-compromised enough to mount an American invasion throughout World War II and is currently Japanese-occupied as a District of the megalopolis of Greater Tokyo. The obscure and interesting sequence of legal wrangling of how it either remained or ended up in Japanese hands is noteworthy.

According to information provided by the Office of War Information, Marcus Island was surrendered by the Japanese to the Americans on August 31, 1945 when the destroyer *Bagley* (DD 386) arrived on the island. There is a very public account by Pharmacist's Mate Joseph M. Clayworth, an eyewitness to the events surrounding this ceremony. His observations, however, are most puzzling. For example, there is no mention at all of the extensive underground facility where all this advanced technology was developed and still existed, nor does he mention submarine access to the island. Further, the territory is described as completely desolate and destroyed. If this was the case, why did the Navy put such an emphasis on destroying it? The Americans, however, were happy to plant their flag and declare themselves victorious. This public account, however, is so typical of what happened through the wind-down period of the war. The Japanese went about their business while the Americans were indulged with face-saving considerations.

Per Article 3 of the *Treaty of San Francisco*, Japan agreed to the United Nations that Marcus Island, amongst several other Pacific islands, would be placed in a trust with the United States as the sole administering authority. All the United States used the island for was to maintain the airstrip and a rather powerful LORAN radio navigation system. It is stated that the purpose of the trust is to ultimately restore these islands to their rightful populations.

The population of Marcus Island was and remains a very vague issue. Officially, there have been no civilians but only Coast Guard and weather station personnel. There were and are, however, extensive operations that take place below the island itself and these are not mentioned.

Immediately after the *Treaty of San Francisco* was signed in 1951, and even before it went into effect on April 28, 1952, a security pact was signed between the United States and Japan. It is officially called, the *Treaty of Mutual Cooperation and Security Between the United States and Japan*. This treaty essentially ensures cooperation between the two countries and positioned the United States to actually protect Marcus Island. Once again, there is no mention at all of the extensive underground facility.

Perhaps the largest testament to Japan having indeed won the war concerns the fact that it regained full and undisputed sovereignty of Marcus Island, for it now holds one of the most valuable resources in today's world of technology.

China has had a virtual stranglehold on the rare-earth minerals market that is used in technology. They have used it to control the world economy. Japan, however, has announced that they have an almost infinite supply of these materials deposited off the shore of Marcus Island. These minerals include ytrrium (used in camera lenses and mobile phone screens, and also europium (Eu), terbium (Tb) and dysprosium (Dy). Aside from being useful, their market value is well over a half trillion dollars. This is just one more instance of Japan finding itself in the driver's seat with regard to world politics and its relationship with the United States.

As superior as Japanese technology was, there were also many political and financial issues that played huge factors as the end of the war played itself out.

ILLUSTRATION OF "FLYING PANCAKES" OVER AN ACTUAL PHOTOGRAPH OF MARCUS ISLAND

CHAPTER TWENTY-SEVEN

THE GOLDEN LILY

In order to understand the politics of 1945, it is absolutely necessary to have a cogent overview of one of the most profound, underestimated, and hidden influences upon world economics and statecraft. This concerns a secret horde of gold that not only changed World War II but still has a significant influence on politics in today's world. Known as the Golden Lily or as Yamashita's Gold, this secret horde of gold is something that most people cannot easily comprehend, not only because of its enormous size, but by the impact it has had on history if only by reason of it being exploited by the Japanese in order to corrupt and turn any American. This horde of gold is famous for never having been turned down. General Douglas MacArthur was no exception, and his relationship with Emperor Hirohito was completely defined and determined by it.

Although certainly not all, a huge amount of this gold horde was acquired as the result of a major event in World War II history which, in spite of appearing in conventional history books, remains under the radar as far as general public consciousness is concerned. Sometimes referred to as Britain's Pearl Harbor, the Japanese launched a massive attack on all British occupied lands in Southeast Asia at the very same time as the attack on Hawaii.

While Emperor Hirohito has been repeatedly portrayed as an innocent marine biologist, in reality, he directed the dilution of the national treasures that were pillaged throughout this large chunk of the world that included most of the islands in the Pacific. This included the wealth of Great Britain, France, and the Netherlands, all of whom had moved their gold to Asia for safety's sake so that the Germans could not take it. In addition to that, there were thirteen other Asian nations invaded by Japan, including French Indo-China, Burma, Malaysia, and the Philippines, the latter being a U.S. territorial possession. This booty included vast amounts of treasures comprising gold, gems, precious metals, coins, art and religious artifacts.

Although this vast horde of treasures are often referred to as Yamashita's gold, most men who were asked about it at the time knew little and this ignorance still applies to this day. In reality, this treasure was Hirohito's. Although Yamashita's name was attached to the treasure, he was merely serving his Emperor. The royal family was put in charge of the entire process and as much booty as was possible was taken back to Japan. The name for the treasure of "Golden Lily" was actually adopted from the name of a poem by Hirohito and used for what was known as the Golden Lily Operation during World War II where the Imperial Japanese Army looted the vast treasures previously described.

As great as this wealth was that was acquired at the onset of World War II, it was all in addition to considerable other riches that had been plundered during the earlier part of the century. This was the result of a very strong wave of Japanese nationalism that arose in response to Commodore Perry's forceful invasion in Japan, the consequence of which was an extreme form of protectionism which extended into the world of organized crime. The most prominent catalyst in this regard was the legendary Mitsuru Toyama, the founder of the Black Ocean Society. In 1895, members of this society literally burned Queen Min of Korea alive in order to destabilize and conquer the Korean Peninsula. This was the same area in which the American railroad barons had wanted to build a railroad that would link Korea and Japan.

Toyama worked with the Yakuza crime faction, and they were fiercely loyal to the Emperor, operating on the basis that they would only carry out their criminal actions in foreign lands and not Japan itself. This served both parties very well. The Yakuza of Toyama worked in conjunction with the Japanese government's secret service. The Queen's assassination provoked an incident justifying an invasion of Korea which became a protectorate of Japan in 1905. What amounted to raw extortion of gold, art and artifacts by organized crime became a justifiable reason to protect Asia from Western influence. All of this, including the activities of the Yakuza crime organization , was done on behalf of the Japanese Emperor.

Initially brought to Japan itself, these treasures were later brought to the Philippines when the Americans started sinking Japanese ships. These treasures were placed under the stewardship of Prince Chichibu, a member of the Imperial Family and the brother of Emperor Hirohito. Prince Chichibu oversaw the excavation of 175 hastily built tunnels all over the Philippines in order to protect the vast crates of wartime loot from where it was to be recovered after the war. There are also vast spaces in Manila that were used by the Spanish, and sometimes their tunnels were modified or constructed in this ancient series of substructures. The general routine was to select a good cave, fill it with treasure and then have party for the workers inside. During the party, the entrance to the cave would be blown up and collapse, thus burying the workers alive. Those killed included thousands of slave laborers and sometimes military men or officials who knew too much.

Although I myself was assigned to burn what I read about these matters, you can now read extensively about what I have summarized above in the works of Sterling and Peggy Seagrave, two American expatriates, now deceased, who published nine books on Asia and Japan and their relations with the West. These include *The Yamato Dynasty: The Secret History of Japan's Imperial Family* (Broadway Books, 2000, ISBN 978-0-7679-0496-4) and *Gold Warriors: America's Secret Recovery of Yamashita's Gold* (Verso, 2003 ISBN 978-1-85984-542-4). Although they did not have all the inside knowledge of the super-dirigibles, Hirohito's acumen for biological warfare, and much of the other knowledge I have shared in this book, they were very much on the mark with regard to the hordes of gold and the clandestine and corrupt liaison between the Japanese and the American hierarchy. They did not know about the enforceable surrender by Truman in terms of suing the Japanese for peace, but all of the damning information they documented and included in their two companion CD-ROM set seals the case for any logical observer.

The Seagraves made a point of the fact that the U.S. still refuses to declassify relevant OSS/CIA materials in blatant contravention of U.S. Freedom of Information laws. They are not, however, privy to the fact that the Government employs experts in documents destruction such as myself. According to the Seagraves, their journey into this morass of golden intrigue began by pure chance when they were commissioned by HarperCollins to write *The Marcos Dynasty* [Harpercollins; 1988] in the early 1980s.

This concerned Ferdinand Marcos, the President and Dictator of the Philippines who was notorious for his corruption, great wealth and brutality inflicted upon his enemies, including the population itself. As Marcos was involved with Yakuza godfather, Yoshio Kodama, who also worked hand-in-hand with the CIA in recovering war loot in the Philippines, the Seagraves had to investigate that aspect of the dictator and his wife's role in laundering golden loot into European, Asian and American banks. Years later, when they were working on *The Yamato Dynasty* and discovered the role played by Prince Chichibu as head of the Golden Lily Operation, they realized they had to reinvestigate the entire legend of Yamashita's Gold from the Japanese point-of-view.

As they wrote their initial books, they met new people, even being contacted by key people with more of the story to share. One of these was Filipino Ben Valmores, the valet of Prince Tsuneyoshi Takeda, the first cousin of Emperor Hirohito. When the burial sites in the Philippines were dynamited and sealed, Valmores was spared. Taking the Seagraves to "Tunnel-8," a massive underground complex of tunnels, he identified many of the Japanese princes involved in its construction. It was empty by that time, however, all the gold had been removed.

Based upon their extensive research, the Seagraves arrived at various reasonable conclusions based upon the many astonishing discoveries they had made. They realized and argued, quite accurately, that Japan was far from the brink of bankruptcy at the end of World War II and concluded that its postwar wealth was consolidated, contained and distributed amongst General Douglas MacArthur and his Japanese and American cronies. While this is a very understandable conclusion, MacArthur was only privy to a relatively small portion of all this immeasurable wealth. Hirohito shared only one tunnel's worth of riches with the General. The wealth of the Emperor was far more vast than even the Seagraves realized. Nevertheless, they understood the magnitude of deception that was perpetrated upon the public, and you can see it clearly in their following statement.

> "Successive U.S. administrations have actually copied Japan by silencing, infantilizing, spoon-feeding and stupefying the public while singing them lullabies of patriotism and moral superiority, and this became easier after 9/11."

The Seagraves were told by their sources that when MacArthur learned of the discovery of the golden horde that he and other important military and government figures had inspected the caves and that President Truman had agreed to keep it secret and off the official books. The funds, under the direction of MacArthur, were put into secret trusts around the world, the most infamous of them all being the Black Eagle Trust or M-Fund (named after U.S. Major General William Marquat, who originally oversaw the fund). The Seagraves take it a step further with detailed corroborations that these trusts were used to bribe Japanese political leaders and later for other priorities of U.S. foreign policy, particularly in regard to fighting communism. Further conclusions of Sterling Seagraves are as follows.

> "The tragedy [for Japan] is that MacArthur handed power back to the same notorious men who started the war, and their so-called Liberal Democratic Party continues to make a joke of democracy today."

> "The most important date in the history of Japan and America is not Pearl Harbor in 1941, or 1945, Occupation in Japan."

"In 1948, the Americans reversed all reforms, halted all punishment of indicted war criminals, rescued Japan's oligarchs, made all records of the war disappear, began freeing everyone from Sugamo Prison (which housed Japan's most infamous war criminals, including top gangsters and psychopaths), and put Japan's government back in the hands of (later Prime Minister) Nobusuke Kishi and other war criminals and drug lords who had conquered all of East and Southeast Asia."

These are all very understandable conclusions based upon their excellent and extensive research. What they did not know, however, was that Emperor Hirohito was acting as the hidden hand, manipulating world events and using General Douglas MacArthur to do so. As I said earlier in this book, the Emperor had the equivalent of one hundred billion U.S. dollars worth of currency in Swiss bank accounts, and that did not even include the hidden gold still buried by reason of the Golden Lily. In such a context, it is ridiculous to either believe or assume that MacArthur had the upper hand.

CHAPTER TWENTY-EIGHT

MacArthur

It was the legacy of the Golden Lily into which General Douglas MacArthur arose to become one of Hirohito's greatest allies as peace was being established between America and Japan, ultimately culminating in the *Treaty of San Francisco* which went into effect on April 28, 1952, the birthday of the Showa Emperor.

It was far from accident or coincidence of fate that MacArthur would be the general who would oversee what is commonly known in the West as the Occupation of Japan. Emperor Hirohito, who had intelligence files on all of his high profile rivals, was fully aware of MacArthur and readily recognized him as being a self-aggrandizing, pompous and egotistical individual who fashioned himself as a nobleman. In actual fact, MacArthur's upbringing and pedigree was as close as one could come to nobility in the United States, such titles being expressly forbidden in the Constitution.

Born into a proud multigenerational military family, Douglas MacArthur's father, Arthur MacArthur Jr., was a Lieutenant General and a Congressional Medal of Honor recipient for his actions as a Union soldier during the American Civil War. Although Douglas MacArthur was severely hazed at West Point, he graduated at the top of his class but not before testifying before a Congressional hearing on hazing at the Academy. He was considered a mama's boy by his fellow cadets, in no small part because his mother moved to Craney's Hotel which overlooks the Academy.

As a young officer, MacArthur was assigned to the Army Corps of Engineers in the Philippines where he worked right alongside his father as an aide. While conducting surveys, he was ambushed by a pair of Filipino brigands or guerrillas and shot them dead, subsequently being promoted. From the Philippines, he then toured Japan and all of Asia conducting inspections with his father. When his father was assigned to Fort Mason in San Francisco, Douglas went with him. From there, he was appointed to be "an aide to assist at White House functions" at the request of President Theodore Roosevelt. All of his postings, particularly the latter, suggest he was the beneficiary of extreme favoritism and it is mentioned here to demonstrate that MacArthur was indeed as close as one might become to being considered American "nobility". He continued working in the Engineering Corps, and when his father died, he was able to arrange to bring his ill mother to Johns Hopkins Hospital. His family heritage and influence was such that he was now appointed to the office of the Chief of Staff.

MacArthur was subsequently transferred to Mexico in 1914 during the United States' Occupation of Veracruz. There, still a member of the Army Corps of Engineers, he was assigned to find some locomotives. This resulted in a fiasco where he was reportedly pursued by various bandidos, shooting and killing several. In what might be determined a very questionable proposal, he was recommended for the Congressional Medal of Honor, but this was denied. Although there were bullet holes in his clothes, one has to wonder, in light of his later behavior and his penchant for shooting brigands while alone, if he was stretching or even creating the truth. There were no serviceman accompanying him, only the three Mexicans he had previously disarmed. In other words, he was the only one of his company who could have shot anyone.

When MacArthur returned to the War Department, he was promoted to major and assigned as head of the Bureau of Information at the office of the Secretary of War, Newton D. Baker. Ever since, MacArthur has been regarded as the Army's first press or public relations officer. Why he was assigned to this role is less important than the fact that he himself, now promoted to colonel, was the antecedent to the Office of War Information. On this job, MacArthur learned how to ingratiate himself with reporters in order to plant stories, spin events to the advantage of the Army and/or himself and to ensure that he was often photographed. It was from this position that he successfully transferred himself to the infantry by literally proposing and creating the 42nd Division, also known as the Rainbow Division, it being named so as it consisted of a collection of National Guard divisions from across 26 states. MacArthur was Chief of Staff of this division.

It was from the 42nd Division, during the Champagne-Marne Offensive during World War I, that MacArthur did seem to genuinely distinguish himself as a military hero and won two Silver Citations (converted later into Silver Stars), and two Distinguished Service Crosses. After the Champagne-Marne Offensive, he won several more Silver Stars and other awards, once again being nominated for and denied a Congressional Medal of Honor. With friends in all the right places, he was promoted to Brigadier General on the day before Armistice Day, the end of World War I. How his experience as a public relations officer played into these awards is not specifically known at this time, but his experience and role as a press officer should not be taken for granted.

After the end of World War I, MacArthur was posted as the Commandant of West Point, a position which enabled him to retain his brevet rank of Brigadier General. At the Academy, he made considerable reforms and improved the overall curriculum in order to groom well-rounded cadets who could respond socially and academically as well as militarily.

Reassigned to the Philippines in 1923, he was eventually promoted to Major General in 1925 upon his return to the United States, thus becoming the Army's youngest and most famous general. In 1930, he was sworn in as Chief of Staff of the United States Army with the rank of General. What is interesting about this period in his life is that he was known to wear a Japanese ceremonial kimono in his office, cool himself with an oriental fan, and smoke cigarettes in a jeweled cigarette holder. During this time, he hired a public relations staff to promote his image with the American public. One of his contemporaries described MacArthur as the greatest actor to ever serve as a U.S Army General. Another wrote that MacArthur had a court rather than a staff.

While all the above gives you an idea of his ability to court the press and his penchant for the Orient, it also gives insight into why Hirohito might have selected him to do his bidding during what is known as the Occupation of Japan. MacArthur's outrageous and hypocritical character traits, however, were in full display at the height of the Great Depression in 1932 when starving veterans of World War I marched on Washington to cash in their bonus certificates for their service in the Great War. The impoverished veterans, who were camping out with their families in protest, at first felt encouraged at the arrival of a show of troops on Pennsylvania Avenue, commanded by MacArthur and Major George Patton. They thought the troops were there to support their cause, but it was quite the opposite. Mac-Arthur and Patton attacked with both calvary and infantry, the latter using fixed bayonets to chase the veterans and their families across Anacostia River to a larger and more secure encampment of protesting veterans. Although President Hoover ordered MacArthur to stop the assault, MacArthur arrogantly ignored him and ordered a new attack, justifying his actions by claiming the march was an attempt to overthrow the Government. As a result of the accompanying tear gas attack, a veteran's wife miscarried and a twelve week old baby died. Sixty people in total were reported as injured. It was a public fiasco.

Major Dwight David Eisenhower, the future president, said about the incident, "I told that dumb son-of-a-bitch not to go down there. I told him it was no place for the Chief of Staff."

When Washington muckraker columnist Drew Pearson and Robert Allen disparaged MacArthur in the *Washington Herald* for his attack on his own veterans, the General sued them for 1.75 million dollars, but it did not stick. They brought forth Isabel Rosario Cooper, a 19-year old Filipino film star MacArthur had brought with him from his last command in Manila and with whom he was having an affair. MacArthur paid her a significant sum, reportedly $10,000 (enough to buy a luxurious house at the time), and the entire matter was closed.

After having served as Chief of Staff of the U.S. Army, MacArthur had reached the peak of whatever authority he might have as a soldier. He then retired in 1937 in order to become a Military Advisor to the Commonwealth Government of the Philippines in what amounted to a very curious and unusual circumstance. Although he was not technically in the U.S. Army, he brought Dwight Eisenhower and an entire staff with him. This was also sanctioned by the President. Before he left for the Philippines, MacArthur convinced the War Department to make an exception to the rule forbidding U.S. officers from receiving compensation from the countries they advised. Thus, whether or not he was technically an army officer while serving as a military adviser, he was breaking what has been called a "rule". There is no question, however, that his role as a military adviser was in the capacity as an employee of the Federal Government.

So that you understand the law concerning such matters, there is what is known as the Foreign Emoluments Clause in the U.S. Constitution, Article I, Section 9, Clause 8 to be specific, that prohibits the Federal Government from granting titles of nobility, and restricts members of the Federal Government from receiving gifts, emoluments, offices or titles from foreign states and monarchies without the consent of the United States Congress. There was never any congressional exception for MacArthur.

The Philippines had been under U.S. occupation since the Spanish American War at the turn of the century. Although the numbers have been deflated by the Office of War Information, the Americans had perpetrated mass genocide against Filipino Muslims during the Spanish American war, killing over three million of them while disparaging them racially. In this respect, the Filipinos saw the United States as oppressors, never forgetting the genocide.

Philippine Commonwealth President Manuel Quezon, who knew MacArthur from his earlier days in Manila, had petitioned the U.S. Government for the General to become his country's top military advisor. The two had a very cozy relationship, and their families were also close. This, however, would change markedly as World War II broke out. Before the war began, MacArthur had insisted on extravagant quarters at the Manila Hotel and even acquired a substantial interest in that as well as the San Miguel Brewery. These interests loomed large on the eve of World War II.

As the threat of Japanese retaliation against the United States economic blockade became imminent and inevitable, President Franklin

Roosevelt recalled MacArthur back to active service. In other words, his role as a military adviser was changed to that as being commander of U.S. Army Forces in the Far East, and it was his job to protect the Philippines against aggression. It was in this capacity that MacArthur made one of the biggest military blunders in American history.

After being notified of the attack on Pearl Harbor, MacArthur did nothing to defend the Philippines or to put planes in the air that could either attack nearby Japanese bases or to defend any incoming Japanese bombers. Ten hours after being notified, he was having his planes being fueled, whereupon they were bombed. This prevented any serious defense of the Philippines. An additional factor working against him was that the war in Europe was prioritized, and only minimal resources were allocated to the Pacific front, most of it being antiquated tanks and other armaments from World War I. MacArthur was in a very tough position, but one of his most unscrupulous characteristics saved him.

After MacArthur was posted as Commander of the Far East in July of 1941, he began to rely on his extensive experience as a press officer and began circulating a significant volume of articles about himself to the American press, lauding himself as the defender of the Philippines and American interests in Asia. All of these articles, idolizing him as America's greatest military hero of the time, were either written or approved by himself. Whatever you might think of this, it ended up saving his reputation if not his very life, as you will soon read.

On Christmas Eve, after Pearl Harbor, MacArthur took the rather astonishing step of declaring Manila an "open city". Just as Hitler had declared Paris an "open city" in 1939, this meant that it was for all intents and purposes non-partisan as to the interests of combatting military forces. In the case of Paris, the Vichy government had taken over and life continued in Paris. All fighting in France was to be done outside of the city. Paris remained intact, for the most part, during the entire war. Manila, however, was an entirely different situation.

As described previously, MacArthur had personal investments in the city of Manila, but historians have exposed that his claim to make it an open city was disingenuous because he most definitely had not removed all of the U.S. military men and installations out of the city. Consequently, the Japanese bombed these targets shortly thereafter. Although they tried to make surgical strikes on just the military targets, there was collateral damage, but it was relatively minimal and the Japanese did what they could to avoid this. In the end, the Filipino people consider the Japanese to be liberators because it was only in 1946 that they gained their independence from the United States.

One of the most condemning facts about MacArthur during this period was only discovered over thirty-five years later in the archives of his aide, Richard Sutherland. It was revealed in 1979 that MacArthur had received a payment of $500,000.00 (this would be about $7,000,000 in today's currency) from his friend, President Quezon of the Philippines. Although this was considered to be payment for his pre-war service, it only happened as a result of MacArthur giving a cold shoulder to his old friend. There are detailed accounts of how it happened, relayed by his stenographer who willingly confessed to having falsified the amount given to MacArthur as only $50,000.00. Payments were also made to MacArthur's staff but one of them was Eisenhower who refused it as it looked improper and would compromise his credibility as an officer of the United States Army. Although apologists have insisted the payments to MacArthur were legal, they were not, and you are referred to the aforementioned Foreign Emoluments Clause in the U.S. Constitution. Although the payments were known to President Roosevelt and the Secretary of War, Henry L. Stimson, they were kept secret and inquiries into them were muffled. This illegal accepting of money in violation of the Emoluments Clause is very relevant to understanding the general circumstances and political power at work that would take place in the wake of Hirohito's biocidal weapons of mass destruction arriving at Tonopah Army Air Field in August of 1945.

As the Japanese began to invade the islands at the beginning of the war, it was the blustery press releases of MacArthur that saved him from further military humiliation and possibly his life. As it became obvious that his command could no longer hold its ground, it was just a matter of time before the Americans would be defeated. Roosevelt thought it would be an extreme public relations blunder not to rescue MacArthur from this indignity because he had been presented to the American public as their greatest general. Accordingly, he transferred MacArthur and his family to Australia to get him out of the fray. It was at this point that MacArthur made his famous statement to the people of the Philippines, saying, "I shall return." What was particularly heinous about the whole affair was that he went to his troops in Bataan and outright lied to them, telling them there were reinforcements and more armaments coming to them. There is no question that he knew this was not the case. The inevitable defeat of his soldiers resulted in the infamous Bataan Death March where his troops, 60,000–80,000 American and Filipino prisoners of war in all, were forced to walk over sixty miles with minimal sustenance resulting in some 600 to 650 American deaths and 5,000 to 18,000 Filipino deaths. It was a complete betrayal of his own troops.

So that you fully comprehend the imperialistic audacity and sheer power that MacArthur wielded, it is highly relevant to cite that, upon returning home from Korea in April 1951 following his dismissal by President Truman, the General was enthusiastically greeted and paraded through the streets by a fawning crowd of a half million people in San Francisco. The following day, he would fly to Washington, D.C. to address a joint session of Congress that would not only test the very strength of the U.S. Constitution but the norms that held it together.* Delivering an unapologetic defense of his policies and a denunciation of his own Executive Commander-in-Chief's (Truman), this was a direct challenge to the concept of civilian control of the armed forces, amounting to an open and direct conflict between elected leader and imperial warlord. The frenzy it produced caused some to fear for the very nature of American democracy itself. In the pandemonium aftermath of MacArthur's speech, Representative Dewey Sort of Missouri shouted: "We heard God speak here today." Herbert Hoover called the General "a reincarnation of Saint Paul into a great general of the Army who came out of the East."

The conservative Senate, however, would deflate the charismatic popularity of MacArthur the demagogue, who otherwise would have been the inevitable candidate for the next presidential election, by conducting an investigation into circumstances of the General's dismissal. The Senate hearings, which went on for weeks, were not favorable to MacArthur who ended up being framed as a warmonger and extremist.

MacArthur's ambition emulated that of Julius Caesar, the latter having crossed the Rubicon and taken over the Roman Republic, only to be subsequently stabbed by the senate. MacArthur did not get that far. His return from across the Pacific represented his own Rubicon, but the Senate nipped his dictatorial aspirations in the bud.

Thus losing public support, the shine of MacArthur's star rapidly dimmed, and the Republicans chose a much more moderate war hero: Dwight David Eisenhower. Had MacArthur but shown a more measured tone during the Senate hearings with appropriately calculated responses, he might well have ascended to the Presidency himself as he was most certainly a more flamboyant and popular personality.

As for the "ever loyal and supporting" Japanese, they were quite taken aback at MacArthur's dismissal. He was, after all, serving them well in keeping the communists at bay in Korea. Deeply affected by the misfortune of his staunch ally, the Emperor himself appeared at the American embassy before MacArthur departed and told him of his

---

*For further information on this tangent, see the third volume of *The Years of Lyndon B. Johnson* by Robert Carol.

distress. Japan's parliament, the Diet, immediately passed a resolution of gratitude in honor of the General.

Regardless of any of the aforementioned misgivings of MacArthur, he was obviously a powerful player on the world stage and had proven himself more than adept in manipulating circumstances in Washington, and he also had extensive experience dealing with Asia. As was alluded to previously, it was the Emperor's recognition of these facts that caused him to demand that MacArthur be his conduit to the West in relation to the post World War II clean up and reconstruction of Japan. None of that could occur, of course, until the Americans finally accepted the terms of the Emperor.

GENERAL DOUGLAS MACARTHUR

CHAPTER TWENTY-NINE

The Surrender

I have already discussed the circumstances surrounding President Truman's decision to drop the atom bombs and why he determined he had to do it, but it was not only for political reasons but also to make him look like a tough guy. This was an image he could readily sell to the American public who never really understood the war in the first place. It was only Truman who could end the war, not Hirohito. All the Emperor could do was use his secret air force to threaten genocide by total annihilation of the United States by delivering biological weapons via huge aero-dreadnoughts that were funded by the looting of all of Asia. Launched from China, these gigantic dirigibles were calculated to reach every part of the continental United States and were to be fortified by the I-400 series of submarines, the Sentoku, the largest in the world prior to the advent of nuclear subs. The Sentoku would unleash medium bombers (the Ki-67 Hiryu "Flying Dragon" which the Americans coded as the "Peggy" bomber) whose fuselages were literally built around the Little Boy uranium ordnance shell developed by the Germans and produced en masse by the Japanese at the Konan facility in North Korea. They were ready, willing and waiting to nuke the Americans from the Eastern Seaboard because the I-400 submarines cruised up and down the Atlantic taking shelter in the massive merchant ships that had been sunk by hundreds by the Third Reich's U-boats. By using those as radar barriers, the Japanese were able to maintain patrol of the Atlantic seaboard from 1945 all the way up to 1951 when the Americans agreed to the peace and all the way up to when the *Treaty of San Francisco* went into effect in 1952.

The public surrender of Harry Truman is on record, but it was spun cleverly so that Americans could not recognize it for what it actually was. After suffering exposure via corruption scandals that rocked his administration at the highest levels and led to the arrest of the majority of the Truman cabinet, the President came on national television after losing the initial conflicts on the Korean Peninsula and admitted to the new reality of "Pax Japanica" to his electorate on September 4, 1951, four days before the *Treaty of San Francisco* was signed on September 8, 1951. In what was the first coast-to-coast television broadcast ever transmitted transcontinentally outside the Third Reich, he personally delivered before all American peoples, in his opening statement to his

Above are rocket parasites attached to the dirigible with one having been launched.

tax paying electorate, that the Japanese had won the war when he made his declaration that a "Peace Treaty with Japan" was deemed effective as recognized by the United Nations organization (of which both the United States, Great Britain and Northern Ireland were founding member states). Truman stressed that all "these nations", including Japan, were equals and that "there must be no victors or vanquished among us".

All of Truman's speech was made in supplication to the 51st birthday-eve day of "Tenji Shama", the first scientist in the history of humanity ever to serve as the head of state, Emperor Hirohito of Japan, the British Field Marshall and English Knight of the Garter.

None of this would have transpired, however, if it were not for Hirohito responding to America's nuclear aggression by unleashing three super-dirigibles from his "Sacred Crane" Task Force and their composited array(s) of parasite attack-craft (including some jet-propelled Horten knock-offs) and their dirigible-coupled super-bombers, all of which (the bombers, fighters, and the disc-dirigibles themselves) were loaded with enough weaponized biocidal contaminants to conceivably exterminate the entire human race (if evenly distributed across the globe).

Launched from the Chinese mainland after the Hiroshima bombing, these reached Nevada three days later, just after the Nagasaki bombing.

THE SURRENDER

ABOVE IS A "HORTON KNOCK-OFF" ATTACHED TO A DIRIGIBLE.

Over one thousand feet in diameter, which is longer than three football fields, these dirigibles crossed the Pacific by reason of the Kuroshio or Japanese wind currents in order to land specifically at Tonopah Army Air Field in Nevada.* The Japanese knew that Tonopah Army Air Field was where the Americans developed their most sophisticated aircraft that included their highest level of technological expertise. Tonopah Army Air Field later became known as Area 51. During the 1940s, the Americans had developed four jets at Homey Air Field as opposed to the hundreds by the Third reich and thousands by the Japanese.

Hirohito, who had already intimidated the Manhattan Project by accurately bombing their nuclear program at Hanford Works with Fu-Go balloon bombs, was now, by deliberately surrendering these huge dirigibles, facilitating his enemy to directly witness the genocidal biological ingredients they carried. After extensive analysis at Fort Detrick in Maryland and knowing too well the lessons of World War I, the American brass knew it was a war-breaker.

The Emperor had delivered the "Showa Ultimatum" and it was now certain that "Operation Olympic" in any form would never go forward; not because it was unnecessary but because it had become unthinkable. This understandably frightened the living hell out of the American generals as well as Truman himself. As was said previously, the President had no choice but to sue the Emperor for peace, all of which

---

*Although three super-dirigibles were dispatched to Tonopah Army Air Field, one of them suffered misfortune and crashed near San Antonio, New Mexico in August of 1945. See Appendix B for more details.

RENDITION OF "FLYING HAMBURGER" DIRIGIBLES LIKE THOSE SENT TO TONOPAH ARMY AIR FIELD

was done on back channels, the paper trail of which I was employed to destroy. Hirohito's request for an acceptable peace, extant for years, was finally accepted but only after he had threatened the Americans with complete annihilation. This is precisely why, when consulted as to what the American public should be told about these heart-stopping new developments in the Japanese-American War, Admiral Ernest J. King boldly stated with flat finality: "Don't tell them anything they don't need to know. When its over, tell them who won."*

Although the Americans had no choice but to pursue peace with Japan, finally accepting Hirohito's demand that he remain the sovereign leader of the Japanese Empire, they nevertheless did an excellent public relations job of spinning the news and treaties to make it look like America had won the war. There are several aspects that demonstrate this beyond any reasonable doubt, but we have all been so brainwashed or misled that it requires some calm thinking.

These dreadnoughts, auxiliary vehicles, and biocides were not the only things the Americans had to deal with. There were also the pilots and crews which consisted of four feet high Yakuza aeronauts, all of whom were in loyal service to the Emperor. They were immediately

---

*Ernest Joseph King (23 November 1878 – 25 June 1956) was Commander in Chief, United States Fleet (COMINCH) and Chief of Naval Operations (CNO) during World War II. As COMINCH-CNO, he directed the United States Navy's operations, planning, and administration and was a member of the Joint Chiefs of Staff. He was the United States Navy's second most senior officer in World War II after Fleet Admiral William D. Leahy, who served as Chief of Staff to the Commander in Chief.

THE SURRENDER

captured and treated as prisoners of war by the Americans. As Yakuza tradition includes the severing of the pinky finger in order to atone for one's sins, their hands had only three to four fingers, a fact which was later used to falsely identify them as extraterrestrials. Although they were prisoners of war, their presence and status has never been officially nor otherwise acknowledged by the United States.

As was said, the biocidal weapons were taken to Fort Detrick in Maryland, and upon analysis, it was determined that if the contents of even one of these dirigibles were distributed evenly across the planet, this could conceivably kill every man, woman, and child on Earth. The biocidal weapons inside the dirigibles were sent by Hirohito in projection of threat as opposed to performing a strike. This was done to induce the Americans into negotiations which would last until September 8, 1951, when a treaty of peace was finally signed between Japan and the United States.

The illustration on the following page is an Italian Manta Bomber shipped to Japan. This was designed like the giant German gliders in order to attack the United States. As gliders, they did not consume a drop of fuel crossing the Pacific because they were attached to the airborne dirigible. Upon arrival, they would have free reign of the United States with an unlimited fuel range as they were primarily powered by the wind currents that had been painstakingly surveyed with the Fu-Go balloon bomb fires. This is essentially the plane that ultimately, like the giant German gliders, was designed to attack the United States. Each one of them carried botulism toxin, one pound of which can kill a billion people.

I have made extensive presentations in the past, showing all of the different sketches of various crafts that were used in the Japanese onslaught. All of the documents concerning these matters, as well as those on the Roswell Incident itself, were brought to the Presidio because this was originally under the province of the Army Air Corps, prior to the creation of the Air Force.

What I am providing herein amounts to FULL DISCLOSURE NOW. No other American has offered this because no other American with access to this information has wanted to let you know that they indeed lost the war and would prefer for you to believe that it is all a matter of extraterrestrials.

Above is a 3-D rendering of the Manta bomber. The ribbed tail is actually three times longer than it appears, but it had to be modified for illustration purposes. Notice the hooking apparatus on the tail for attaching to a dirigible.

CHAPTER THIRTY

The Tribunals

As was previously alluded to, part of the back channel negotiations that secured the initial peace between Japan and the United States included the accedence to Hirohito's demand that General MacArthur would serve as the Emperor's intermediary with regard to all American operations in Japan. There were several reasons for this; and besides MacArthur's penchant for nobility and aggrandizement, the latter making him susceptible to outright bribery, he was also somebody who was adept at ordering people about and actually getting things done. He was a very useful servant of the Emperor's agenda; and all things considered, he served him well. This can best be understood in the war crimes trials that purged those in his Japanese military who were treasonous to the Emperor, all the time being carried out by willing Americans. This enabled Hirohito to avoid looking like a ruthless Josef Stalin who was purging his own enemies, but it also gave the Americans satisfaction that they were punishing vile foes. The press generally criticized MacArthur's "show trials" as they were often referred to as they clearly did not pursue the true perpetrators on the Japanese side, including the Emperor himself.

Although MacArthur's special proclamation ordering the establishment of an International Military Tribunal for the Far East was on January 19, 1946, he took care of General Tomoyuki Yamashita before all of that in the preceding months in what amounted to a true mockery of justice that was unprecedented, even for a wartime military court. While MacArthur's tribunal has been severely criticized by the press for a number of reasons, the gravest and most outrageous example is the prosecution of General Yamashita, the man who was responsible for burying all the gold.

In a regular criminal trial or military court martial, there are certain rules that are followed that are very specific with regard to evidence and the accused's rights to defend himself. In the case of Yamashita, it was a different type of trial that followed neither criminal law nor the standard procedures for a court martial. You can read about this independently if you want as it has been written about extensively and is heavily criticized. The press reported the procedures to be outside the norms of "Anglo-Saxon justice". I will only give a brief summation.

Basically, Yamashita was found guilty and executed by reason of the fact that his troops savagely killed people en masse, raped women and committed all kinds of atrocities and horrendous war crimes. Although no one denied this, there was no evidence at all that Yamashita had either ordered it or was even aware of it. In fact, the American officers assigned to be his defense attorneys were initially hostile to him by reason of the enmity of the war itself. After meeting him, hearing his story and examining all available evidence, they concluded that he was indeed not guilty. The magistrate, however, was bored by the proceeding, told them to hurry it up and just had no appetite for judicial proceedings. Instead, he did MacArthur's bidding and convicted Yamashita, sentencing him to be executed. There is, however, a significant backstory which is not realized by practically anyone.

Yamashita was previously implicated in a coup against the Emperor in 1936. Realizing he had fallen into disfavor with the Emperor, he wanted to resign from the Army but his superiors forbade it. Yamashita was eventually assigned to missions of plundering and hiding gold for the Emperor. When it came time to secure the secrecy of the locations, it was expedient for the Emperor to have Yamashita executed. As far as the Japanese public were concerned, however, MacArthur was the bad guy and took the blame.

In spite of his treachery towards the Emperor, Yamashita managed to salvage his own soul, at least from the perspective of many. After he was initially exiled to a post in Korea after his disloyalty, he studied Zen Buddhism and was said to have benefited. At his trial, which has been deemed racist in the extreme, he praised and complimented his judges and the American officers who were executing him. In other words, he showed extreme graciousness. The reason for this is that a deal had been made. If he would accept his fate as a gentleman, the Americans told him that the United States would release all the Japanese-American citizens they had interred on the onset of the war. Even though most of these citizens were released from detention camps, they were not given back their citizenship and only a very small percentage remained in America. Most returned to Japan or emigrated to Brazil.

Yamashita's trial established a new phrase in military jurisprudence which is known as the "Yamashita Standard". This means that a commander is responsible for any of the criminal actions of his troops whether or not he is aware of them or whether or not he is directly or indirectly involved. This contrasts to what is known as the "Medina standard" which is based upon the 1971 prosecution of U.S. Army Captain Ernest Medina for the infamous My Lai Massacre in Vietnam. Although he was reported to have ordered the killing of women and children, i.e. any who were "walking, crawling or growling", Medina was acquitted.

Most of the trials of MacArthur's International Military Tribunal followed the unjust procedures of the trial of Yamashita. These show trials frustrated the American press and public because they were obviously farcical and did not prosecute the primary perpetrators of the war. Although many knew Hirohito was responsible, few realized the degree to which he was involved or how he was the actual engineer behind the Japanese war machine. Common history demonstrates that MacArthur fiercely defended the Emperor, and he did it in such a way that the public, aided by the Office of War Information, has bought it. This is why MacArthur developed the moniker of "the Emperor's General".

Both the Emperor and the General served each other very well. The public has remained in the dark. Hirohito allowed the Americans to orchestrate whatever public relations they wanted, but he also got what he wanted: an end to hostilities and an open market for Japan. His country prospered as never before. He also succeeded in preventing the United States from ever again declaring war.

While these issues have been woven into the public consciousness in a way that suits the military, the true facts of the history of the matter leave a clear trail telling you that what I offer here is true. No matter how one might attempt to manipulate or spin what is actually stated, one of the most conclusive testaments to the Japanese victory is in Article 14 of the *Peace Treaty of San Francisco*, as follows:

> "It is recognized that Japan should pay reparations to the Allied Powers for the damage and suffering caused by it during the war. Nevertheless it is also recognized that the resources of Japan are not presently sufficient... the Allied Powers waive all reparations claims of the Allied Powers and their nationals arising out of any actions taken by Japan."

By signing the treaty, Allied countries waived all rights to any claims, including claims by their citizens and servicemen forced into slave labor by the Japanese warlords. Besides all of that, none of the assets of the Golden Lily were ever considered in regards to potential reparations. The Japanese Empire, and the Emperor in particular, had retained all of his virtually unlimited assets.

THE ROSWELL DECEPTION

CHAPTER THIRTY-ONE

THE ROSWELL INCIDENT

There are multiple threads of events and circumstances which led to and precipitated the so-called Roswell Incident, popularly described as the crash of a flying saucer resulting in the discovery of dead aliens and a monumental cover-up by the Government.

In this book, you have already been familiarized with the initial thread: Emperor Hirohito's background with regard to being an internationally renown doctor of marine biology and the richest man in the world responding to the genocidal mania articulated by the American racist, Jack London, by developing sophisticated biocidal weapons of mass destruction which were unleashed upon China initially and eventually presented to the American high command in a show of strength so as to eventually bring an end to World War II.

The second thread would be the American response to the looming Japanese threat during the Battle of Los Angeles. This consisted of an all-out panic and the self-destructive response of lethally vaccinating their own soldiers followed by a disinformation campaign suggesting that the flying craft over Los Angeles were extraterrestrials. The alien theme set up further disinformation campaigns designed to obfuscate the vulnerability, incompetence and treachery of America's military.

Another very important thread which I have not yet touched upon and can only briefly summarize in this book has to do with the German flying craft that were developed prior to and during World War II. The German flying craft or discs are well documented by many different sources, and it is not the purpose of this book to convince you that there was such a program.* What is important here is that it played an extensive role in what eventually became known as the Roswell Incident.

---

*The first American publication to feature photos of the German flying saucers was *The Black Sun - Montauk's Nazi-Tibetan Connection* by Peter Moon, published by Sky Books. The photographs in *The Black Sun* first appeared in both the German and English versions of the German publication *Secret Societies and Their Power in the Twentieth Century* by Jan Van Helsing. Soon after the fall of the Berlin Wall, these photos were supplied to the author by members of the Knights Templar who had lived in East Germany. Prior to these publications, David Childress of Adventures Unlimited released an extensive video on German Flying Saucers with plenty of actual footage of the craft. An even earlier researcher on the topic was Henry Stevens of the German Research Project who collected extensive documents on the flying craft that included illustrations and diagrams. Stevens eventually published a book with Adventures Unlimited entitled *Hitler's Flying Saucers: A Guide to German Flying Discs of the Second World War*.

German Flying Craft

Over a year before Germany was occupied, the Third Reich began a massive exodus of technology, personnel, and resources to the Southern Hemisphere and particularly Antarctica. Extensive exploration of the continent had begun in 1938 when Hitler decided to conduct the most comprehensive scientific study of Antarctica that the world had ever known. The expedition was conducted by Captain Alfred Ritscher under the orders of Hermann Göring. Aboard a homemade aircraft carrier, Ritscher sailed a fleet of ships off the coast of the area known as Queen Maud Land. During two expeditions, his planes photographed over 100,000 square miles and claimed over twice that amount of territory. The borders were marked with swastika pennants which had been attached to javelins so that they would stick in the ice and remain upright. They called their new continent Neuschwabenland. The stated military purpose of the mission was a ruse from the very start. The Germans claimed to be studying the feasibility of whaling, but this was ridiculous. The Norwegians had been successfully whaling in the area for years. There were different reasons for the expedition itself. While it provided a safe base in the event of defeat on the European continent, it also provided a safe operational environment for an extensive computer system which was far ahead of its time. There was also an agenda to link to "Unterland", the inhabitable region beneath the Antarctic snow.

Few genuine public comments were ever made about this secretive mission. One of the more remarkable quotes came from Grand Admiral Karl Dönitz, the Commander in Chief of the German Navy in 1945 and Hitler's eventual successor, when he said, "The German submarine fleet is proud of having built for the Führer, in another part of the world, a Shangri-La on land, an impregnable fortress." This statement was made in 1943 and reported in the *National Police Gazette* in 1977. On the following page is a partial schematic of the German base in "Unterland".

The Third Reich's exodus to Antarctica towards the end of the war included flying craft that were saucer-like in shape which also had weapons. While these had not been developed to their fullest extent in the years leading up to the conquest of Berlin, they were too little too late to prevent the occupation of Germany. Their capabilities, however, vastly improved as the war was coming to a close in Europe and particularly afterwards.

As was stated in Chapter 3, the duly elected government of the German people, the National Socialists, never surrendered to the Allies. It was only the military that surrendered. Accordingly, the National Socialists continued to wage war against the Allies and especially the United States. In fact, there was a war of huge proportions going on in

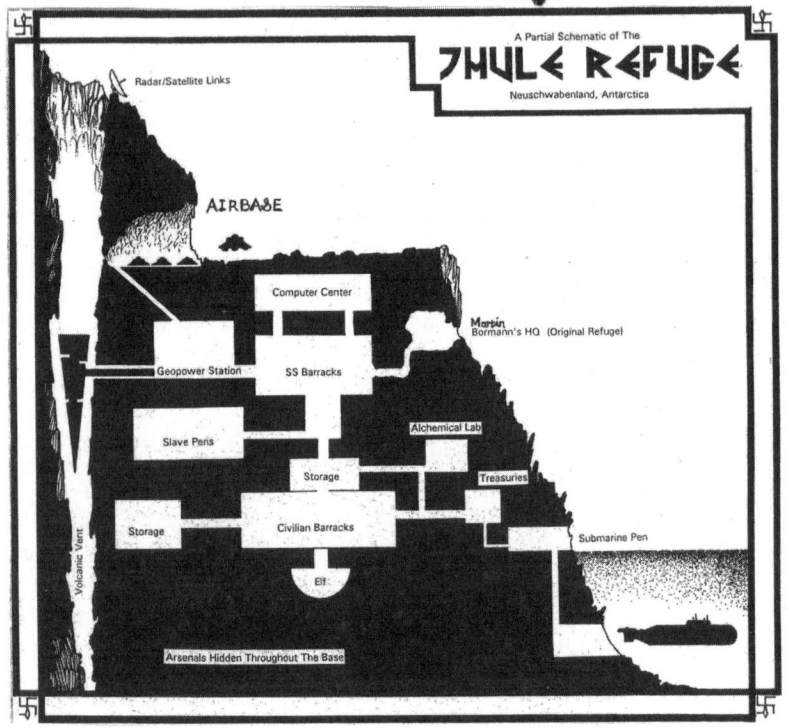

PARTIAL SCHEMATIC OF THE GERMAN BASE IN "UNTERLAND"

the skies, and this continued from 1945 and even past the date of when the *Treaty of San Francisco* went into full effect on April 28, 1952.

There are at least two books which cover the subject of "disc wars" between the Third Reich and the U.S. Government, both of which used titles excerpted from the singularly salient standard order: "Shoot Them Down By Any Means Necessary", the latter being a direct extension of Franklin Delano Roosevelt's Executive Policy of Hostility against multiple so-called "Alien Races":

> *By Any Means Necessary: America's Secret Air War* by William E. Burrows (2001) ISBN 978-0374117474 Farrar, Straus and Giroux and
> *Shoot Them Down! The Flying Saucer Air Wars of 1952* by Frank Feshino, Jr., (2007) ISBN 978-0615155531

The United States was warring against the Communist Bloc and the so-called UFO presence simultaneously with 166 crewmen acknowledged as having been killed while probing Russian and Chinese defenses. Between 1950-1959, there were also thirty cases of intercepts of scrambled jets on Combat Air Patrol guarding the Washington D.C. perimeter whose mission orientation was targeting UFOs.

Besides warring against the flying discs, the Americans did everything they could to understand and reverse engineer the craft they were able to shoot down. What is important in this work, however, is not to document all the instances or even the fact that it did occur. That is offered here as a "given" and you can do your own homework on the subject if you wish. Once again, it is an extensive subject, and there are also several youTube videos covering the topic. What is important here is the role that these issues played in the Roswell Incident.

At the very beginning of the saucer wars and what promulgated their continuance was Admiral Richard Byrd's well publicized trip to Antarctica in 1946, promoted to be a geographical survey of the land. Although Byrd (USN, Retired, Officer in Charge, Task Force 68) organized the expedition, it was led by Rear Admiral Richard H. Cruzen, USN, Commanding Officer, Task Force 68. Titled Operation HIGHJUMP, this purported scientific mission included many thousands of troops, a battleship, an aircraft carrier, at least thirteen ships and 33 aircraft as well as an entire support team. All of this was far more than a geophysical investigative team required. Admiral Byrd, who was an explorer rather than a military leader, was taken along as a figurehead in a shell operation. The United States was seeking to disperse and destroy any remnants of a Nazi safe haven in the Antarctic as previously alluded to in the statement of Admiral Karl Dönitz.

Officially commencing in August of 1946, the fleet did not reach the Antarctic until very late December. Expecting to remain in the Antarctic region for eight months, their trip was cut short from eight months to about eight weeks, ending in February of 1947. What defeated the U.S. Navy was radar, and it was a radar the Japanese had developed and shared with the Germans. The radar was coordinated into such a system that the German contingent could wipe out Marine landing units, and this is why you never hear about the thousands of marines who were killed. You hear about a few sailors, but the marines who were flown in on Douglas Boxcars from New Zealand were wiped out to the last man because of this radar response system. Like the dirigibles of World War One, the Japanese dirigibles were designed to fly above the radar line so that they could release parasite crafts that could thereafter drop bombs after having swept in below the radar line without being observed. This stopped the American forces dead in their tracks. The U.S. Navy returned with its tail between its legs.

Task Force 68 had been a Naval and Marine Corps effort to dislodge the Third Reich's sanctuary in Antarctica. As a result of their failure, the Government turned the objective over to the Army. It was the Army who had confirmed the genocidal capacity of the biocidal weapons contained in the Japanese dirigibles that had landed in the jurisdiction of Tonopah Army Air Field. The dirigibles remained in the possession of the Army.

As the Navy's planes, ships and troops had been stifled, it was the Army's "brilliant" idea to use these gifted super-sophisticated dirigibles to carry a Fat Man plutonium bomb in one of the parasite bombers attached to the dirigible. This way, they could fly it above the radar line and blast the hell out of the Third Reich's compound in Antarctica, thus sealing the various entrances to "Unterland" and putting a stop to the German exodus as well as an end to the disc wars which had only just begun. Flying the dirigibles, however, was a problem to the Americans because they not only did not know how

ALTHOUGH IT IS CONTINUOUSLY DENIED OR UNREALIZED, THE JAPANESE WERE AND ARE ROUTINELY FAR AHEAD OF THE REST OF THE WORLD WHEN IT COMES TO TECHNOLOGY. ABOVE IS AN EARLY ACOUSTICAL AIRPLANE WARNING SYSTEM DEVISED BY THE JAPANESE IN THE 1920S. IT WOULD PICK UP THE BUZZ OR HUM OF AN INCOMING PLANE.

THE ROSWELL INCIDENT

Figure 3. Navy radar Mark 1 Model 1 Type 2 Designation 11. Daito Naval Experimental Facility Collection. Courtesy of Kawamura Yutaka.

Figure 4. Navy radar Mark 1 Model 3 Type 3 Designation 13. Daito Naval Experimental Facility Collection. Courtesy of Kawamura Yutaka.

ABOVE ARE JAPANESE RADAR UNITS, ANOTHER EXAMPLE OF THEIR TECHNOLOGICAL ADVANCEMENT DURING THE WAR YEARS.

THE ROSWELL DECEPTION

Figure 7. Army radar Tachi 24 (Würzburg Type). Courtesy of the U.S. National Archives (331-SCAP-30A).

Figure 8. Japanese magnetron. All metal type. Daito Naval Experimental Facility Collection. Courtesy of Kawamura Yutaka.

ABOVE IS ANOTHER EXAMPLE OF JAPANESE
RADAR AND ALSO A MAGNETRON.

THE ROSWELL INCIDENT

to fly any of the Japanese flying craft, they were physically too big to get inside and operate what the Japanese had rendered to them. It was not a problem, however, for the Yakuza crew who were small in stature, not being greater than four feet in height.

In order to carry out their intention to bomb Antarctica, the Americans tortured the uncooperative Yakuza crew who were nothing more than prisoners of war, and this was in spite of the fact that the Office of War Information was declaring that the war was over after the so-called "Unconditional Surrender" aboard the *USS Missouri*. They not only wanted the Yakuza to explain to them how to operate the dirigibles and the other attachable aircraft such as the Manta Bombers, they forced them to experiment with them on behalf of an American agenda.

There were various experimental flights that took place before the Roswell Incident. One of these resulted in the now famous accounts of Kenneth Arnold spotting a series of UFOs near Mount Rainer. Arnold was flying a small private plane when he allegedly clocked the UFOs at an estimated speed of twelve hundred miles per hour on June 24,

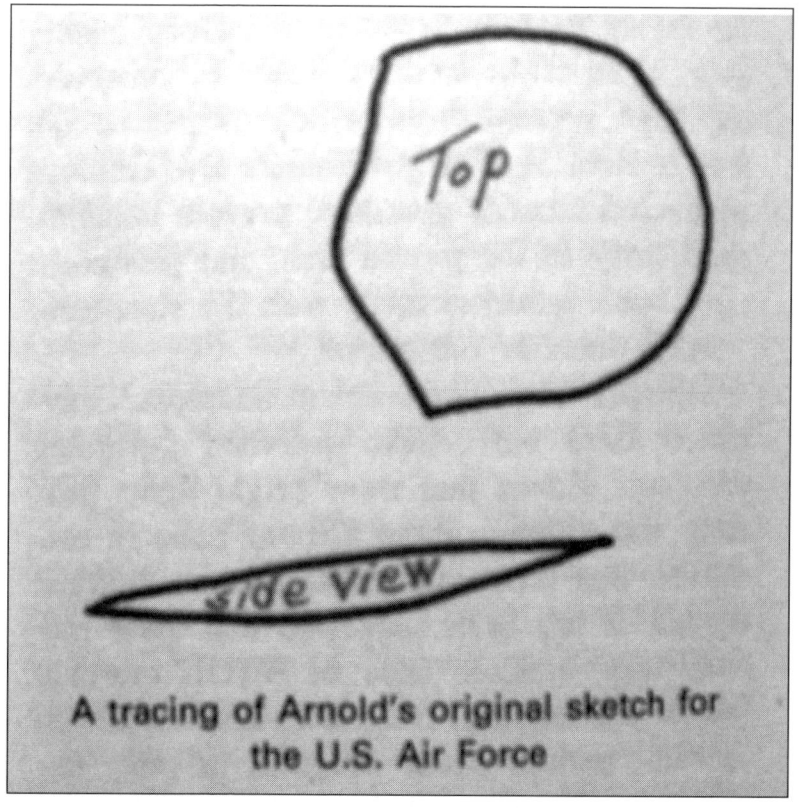

A tracing of Arnold's original sketch for the U.S. Air Force

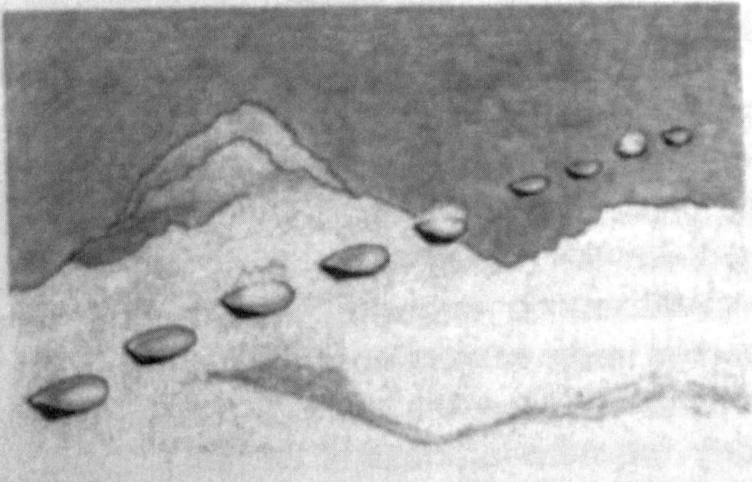

A re-creation of the scene according to Arnold's description (Drawing by Susan Swiatek)

ABOVE IS THE AMERICAN VOUGHT V-173 FLYING PANCAKE.

THE ROSWELL INCIDENT

THE ABOVE RENDERING IS A STREAMLINED JAPANESE VERSION OF THE V-173 WHEREIN THE PILOT REMAINED PRONE WHILE FLYING. NOTICE THE HOOK ON THE EXTENSION FOR ATTACHING TO THE DIRIGIBLE.

1947. This sighting was significant because it is the first sighting to get coverage from national news outlets, and it also introduced the term "flying saucer" into the popular lexicon. It occurred two weeks prior to the Roswell Incident itself.

What Kenneth Arnold observed and reported were disarmed Vought V-173 Flying Pancakes, also known as a "Zimmer's Skimmer", which had been modified by the Japanese after having purchased the blueprints. These were redesigned so that the aircraft could be suspended from the super-dirigibles, but the modifications also enabled maximum aerodynamic efficiency by having the pilot fly prone so that the contour of the fuselage was optimally streamlined. To the tiny Yakuza, this position was not objectionable and is by far the most comfortable in case of airsickness incurred through the violent aerial acrobatics that are demanded in combat. The very embodiment of aerial agility, all nine of these aircraft were originally painted with the images of windblown leaves over the entire fuselage along with their Japanese national colors, but the Americans had painted over this in turn with a highly reflective silver metallic sheen.

Twin engines were used in these modified aircraft so that torque and gyroscopic couples would be neutralized to provide structural and aerodynamic efficiency as well as to make possible the continuation of flight on one engine in case of failure of the other. Both engines were light weight per their horsepower and were connected to the propellers by means of conventional reduction gearing and clutches whereby, if one engine stopped, it might be disconnected from its propeller so that the two propellers would be driven by the other engine at sufficient speed to enable the pilot to return to the mothership and bring the machine in safely.

Another feature of novelty which increased overall efficiency above that of other aircraft at the same speed is the location of the propellers at the "wingtips" and their rotation in such directions that most of the energy which would otherwise be lost in the twist of the slipstream would be returned to the machine in the form of diminished induced drag.

The ultimate outcome of all this ingeniously unique and efficient aerodynamic engineering resulted in Kenneth Arnold seeing a group of nine Flying Pancakes flown by prisoners of war who were, much to the chagrin of the Army, making a break from their captors. These Yakuza were from the criminal class of Japan, the "Samurai Mafia" who had undergone the cultic initiation ritual of removing their pinkies and sometimes ring fingers as well, leaving the appearance of having only three or four fingers. Accordingly, the cockpits and controls of the various craft they flew were not only designed to suit four feet tall men but those with as few as three fingers. To the degree any Americans could conceivably fit into any of the aircraft, the controls were not suitable for them to operate.

OBJECT "SEEN" BY KENNETH ARNOLD AND WHAT THE AMERICAN MILITARY MADE HIM HOLD UP AS THEY WANTED HIM TO SAY, "I SAW THIS."

Following is the text of the newspaper article that appears in the captioned photo of Kenneth Arnold (on the opposite page):

PAGE 2 *THE CHICAGO SUN*, THURSDAY, JUNE 26, 1947
In These United States

Supersonic Flying Saucers Sighted by Idaho Pilot

Speed Estimated at 1,200 Miles an Hour When Seen 10,000 Feet Up Near Mt. Rainier

PENDLETON, Ore, June 25.

Nine [bright], saucer-like objects flying at "incredible" speed at 10,000 feet altitude were reported here today by Kenneth Arnold, [Boise] (Idaho) pilot, who said he could not hazard a guess as to what they were. Arnold, a U.S. Forest Service employee searching for a [missing] plane, said he sighted the mystery craft at 3 p.m. They were flying between Mount Rainier and Mount Adams, in Washington state, he said, and appeared to weave in and out of formation. Arnold said he clocked them and estimated their speed at 1,200 miles an hour. Inquiries at Yakima last night brought only blank stares, he said, but he added he talked today with an unidentified man from [Ukiah], south of here, who said he had seen similar objects over the mountains near [Ukiah] yesterday. "It seems impossible," Arnold said, "but there it is."

Just as the Flying Pancakes had been painted, so did the Americans paint over the Japanese colors of the dirigible with highly flammable silver nylon paint with a reflective capacity that would make it shine in the moonlight and be visible under cover of darkness so that the Americans could visually keep track of it.

The Yakuza were directed to flight-test these craft at night and they tried to box them into a confined area of airspace in the most isolated area of the United States that was in the Four Corners region. They kept it within a coordinated air space between escorts, trying to make them demonstrate the dropping of a dummy bomb. As is common in that area, a sudden thunderstorm came up, and after winds had blown the craft towards Roswell, the Yakuza opened up all the vents to the hydrogen peroxide gas cells in their aero-dreadnought. As the balloons lifted with hydrogen peroxide, they were highly flammable. The Americans, who could not fit into the confined spaces of the dirigible, were completely unaware that the Japanese POWs had hidden flares in the cockpit. Blowing their flares into the hydrogen peroxide gas cells, the Yakuza took the opportunity to self-immolate rather than be subjected to continued imprisonment and torture by their American captors. They literally blew the massive aero-dreadnought out of the sky. The result was that you had between an estimated 600-1,000 auditory witnesses to that huge explosion. That is how loud it was.

When it happened, you had people who saw the burning of the discharged craft. Material was flying and landing all over the place with the main craft itself having its own separate crash area. There was also the parasite craft which was linked onto the plane with a catapult system and designed to discharge the bomb as it drops. While only a dummy bomb was being tested in this case, it was itself another egg-shaped oval "craft" that was found in another place. This is why you had all those separate instances and up to as many as three crash sites with debris strewn hundreds of yards across the New Mexico desert. This is also why you have an idiotic assertion by Stanton Friedman (more about him later) who stated that two alien craft had flown into each other and blew up; hence, two separate crashes, as if they were playing "chicken" with each other after having flown from faraway star systems. Although this is a ludicrously absurd proposition, people just nod their heads as if it all makes sense that aliens from an advanced civilization were flying while intoxicated or that their detection systems simply did not work. While it is true that the civilization who created that craft was thousands of years old and far more advanced than that of the Americans, it was not an alien civilization in the sense that most Americans think. At that time, however, the Japanese were an enemy alien civilization in the context of a foreign country that was still very much indeed at war against the United States.

The huge explosion of the dirigible and its attachments propelled itself in many directions, and this is why William Woody and his father, some of the most important witnesses at Roswell, saw the Manta bomber going down through the sky. They said that it was too slow to be a meteorite. It was a plane on fire. They drove over to it and ran into a military cordon twenty miles long. There is no way in hell that aliens could spontaneously crash and get a military cordon twenty miles long in place before such a crash occurred. Accordingly, William Woody and his father are never cited in the usual stories of Roswell witnesses. You do, however, get other accounts whose edited comments fit the desired narrative of the military itself.

There was also the matter of the Yakuza. Some of the crew actually survived the conflagration, and this is why at every one of these sites in Roswell, you had dead bodies and survivors. The majority

THE ROSWELL INCIDENT

BELOW ARE GRAPHIC REPLICATIONS BY JESSE MARCEL OF THE MARKINGS ON THE DEBRIS FROM THE ROSWELL CRASH.

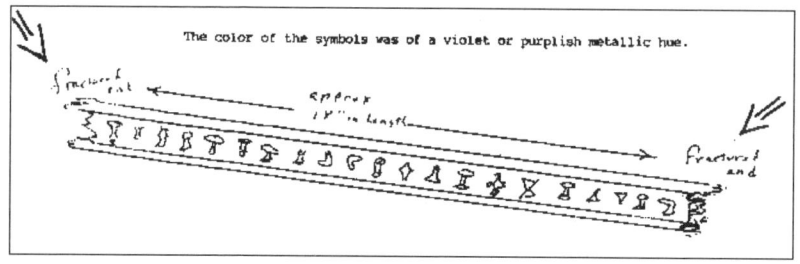

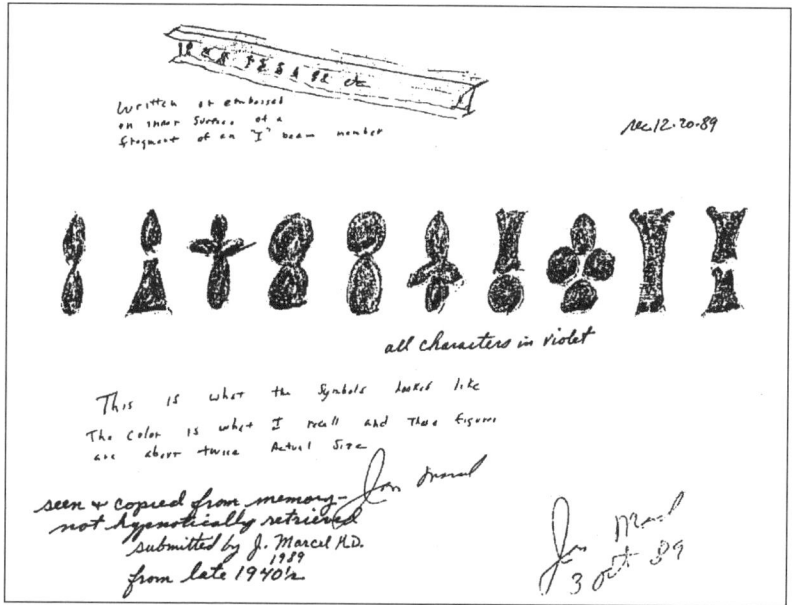

were killed, but some survived, eventually succumbing as a result of the beatings that accompanied their interrogations. All of the POWs were burned, and although some early accounts have them identified as being clearly Japanese or Oriental, the bodies have been frequently described as aliens with a yellowish-orange or gray skin color. All of them were indeed Yakuza and the skin color was attributable to severe burns and then rigor mortis having set in.

Another aspect to the Yakuza which made it tempting to either view or portray them as aliens was that, due to the static electricity that they would have experienced while crossing the Pacific, they would shave all of their body hair to ensure that they would not get electrocuted to death. This included shaving their eyebrows and pubic hair. The only thing they would not shave would be their eyelashes. Their short stature explains why the coroner was ordered to provide small coffins for the "aliens". There is also the attribute of them having three fingers, and the reason for this has already been stated.

Accounts of a so-called retractable metal have also reinforced in people's minds that this was indeed an alien ship that had crashed. This material was not metal at all but was rubberized silk that was the outer coating of the balloons, a substance that Americans were and are completely unfamiliar with. The strange hieroglyphics you read or hear about were nothing more than the Yakuza's stencil renditions of their culture's ideogrammatic special characters that they used to identify their controls or systems.

The Army had a potential mess on their hands with regard to public relations. They had multiple crash fields with all sorts of debris strewn across a vast area plus living and dead bodies requiring an inordinate amount of explanation. If truthfully exposed and accounted for, however, it would reveal the Army to be guilty of war crimes by reason of keeping these soldiers as prisoners of war and torturing them to the point of suicide. There was also the prospect that America's story of victory over the Japanese could ultimately be undermined. They had to act fast, and so they did. Completely dedicated to obfuscation of the truth and spinning a story that would distract the public, it was convenient and expedient for them to make a press release announcing that this crashed aircraft was a flying saucer. Ever since, this myth has been a closely cherished staple of the UFO community and the American public in general.

CHAPTER THIRTY-TWO

THE COVER-UP

The magnitude of the lies and disinformation that has been unleashed upon the American public with regards to World War II and Roswell is so enormous that it cannot be easily fathomed by most anyone, and the implications and unsaid aspects run yet even deeper into the souls of all men. The purpose of this book is to summarize and expose key falsehoods surrounding these subjects so that the general public can dispel themselves of the primary false notions that have been fostered upon them by the Office of War Information and its successors. Thousands of sets of documents that I collated or destroyed clearly demonstrated beyond any shadow of reasonable doubt that the Roswell crashes constituted a collateral damage incident of the Second World War, an incident that threatened to expose the fact that the Greater Pacific War was still being waged at that time, to America's loss.

In dealing with the Roswell crash, the Army was consistent with their experience and public relations strategy from the Battle of Los Angeles by immediately issuing a press release stating that a flying disc had crashed in the New Mexico desert. It is quoted below.

> "The many rumors regarding the flying disc became a reality yesterday when the intelligence office of the 509th Bomb Group of the Eighth Air Force, Roswell Army Air Field, was fortunate enough to gain possession of a disc through the cooperation of one of the local ranchers and the sheriff's office of Chaves County. The flying object landed on a ranch near Roswell sometime last week. Not having phone facilities, the rancher stored the disc until such time as he was able to contact the sheriff's office, who in turn notified Maj. Jesse A. Marcel of the 509th Bomb Group Intelligence Office. Action was immediately taken and the disc was picked up at the rancher's home. It was inspected at the Roswell Army Air Field and subsequently loaned by Major Marcel to higher headquarters."

The *Roswell Daily Record* of July 8, 1947 immediately headlined the story "RAAF Captures Flying Saucer on Ranch in Roswell Region", providing a historical artifact that, in retrospect,

seems ready-made for fueling an episode of *The X-Files*. See the headlines above and a replication of the words in the article here:

(HEADLINE) "RAAF Captures Flying Saucer on Ranch in Roswell Region"
(SUB-HEADLINE) "No Details of Flying Disk Are Revealed"
(SECOND SUB-HEADLINE) "Roswell Hardware Man and Wife Report Disk Seen"

"The intelligence office of the 509th Bombardment group at Roswell Army Air Field announced at noon today, that the field has come into possession of a flying saucer.

According to information released by the department, over authority of Maj. J. A. Marcel, intelligence officer, the disk was recovered on a ranch in the Roswell vicinity, after an unidentified rancher had notified Sheriff Geo. Wilcox, here, that he had found the instrument on his premises. Major Marcel and a detail from his department went to the ranch and recovered the disk, it was stated.

After the intelligence officer here had inspected the instrument it was flown to higher headquarters.

The intelligence office stated that no details of the saucer's construction or its appearance had been revealed.

Mr. and Mrs. Dan Wilmot apparently were the only persons in Roswell who seen (saw) what they thought was a flying disk.

They were sitting on their porch at 105 South Penn. last Wednesday night at about ten o'clock when a large glowing object zoomed out of the sky from the southeast, going in a northwesterly direction at a high rate of speed.

Wilmot called Mrs. Wilmot's attention to it and both ran down into the yard to watch. It was in sight less then a minute, perhaps 40 or 50 seconds, Wilmot estimated.

Wilmot said that it appeared to him to be about 1,500 feet high and going fast. He estimated between 400 and 500 miles per hour.

In appearance it looked oval in shape like two inverted saucers, faced mouth to mouth, or like two old type washbowls placed, together in the same fashion. The entire body glowed as though light were showing through from inside, though not like it would inside, though not like it would be if a light were merely underneath.

From where he stood Wilmot said that the object looked to be about 5 feet in size, and making allowance for the distance it was from town he figured that it must have been 15 to 20 feet in diameter, though this was just a guess. Wilmot said that he heard no sound but that Mrs. Wilmot said she heard a swishing sound for a very short time.

The object came into view from the southeast and disappeared over the treetops in the general vicinity of six mile hill.

Wilmot, who is one of the most respected and reliable citizens in town, kept the story to himself hoping that someone else would come out and tell about having seen one, but finally today decided that he would go ahead and tell about it. The announcement that the RAAF was in possession of one came only a few minutes after he decided to release the details of what he had seen."

The Army knew damn well it was not a flying saucer in the sense of it being an extraterrestrial craft, but they deliberately planted a seed. The next day, however, they spun the story to a different tune, claiming it was a weather balloon. Keep in mind that when the press used the words "flying saucer", this term had just been coined by the Kenneth Arnold sighting weeks earlier. This was a consistent meme planted in the minds of a gullible public whose gullibility would be exploited and exacerbated for more than seven decades. The following quotation is from the *Roswell Daily Record* of July 9, 1947.

(HEADLINE) **"Harassed Rancher Who Located 'Saucer' Sorry He Told About It"**

"W. W. Brazel, 48, Lincoln county rancher living 30 miles south of Corona, today told his story of finding what the army at first described as a flying disk, but the publicity which attended his find caused him to add that if he ever found anything else short of a bomb, he sure wasn't going to say anything about it.

Brazel was brought here late yesterday by W. E. Whitmore, of radio station KGFL, had his picture taken and gave an interview to the Record and Jason Kellahin, sent here from the Albuquerque bureau of the Associated Press to cover the story. The picture he posed for was sent out over AP telephoto wire sending machine specially set up in the Record office by R. D. Adair, AP wire chief sent here from Albuquerque for the sole purpose of getting out his picture and that of sheriff George Wilcox, to whom Brazel originally gave the information of his find.

Brazel related that on June 14 he and an 8-year old son, Vernon, were about 7 or 8 miles from the ranch house of the J. B. Foster ranch, which he operates, when they came upon a large area of bright wreckage made up on rubber strips, tinfoil, a rather tough paper and sticks.

At the time Brazel was in a hurry to get his round made and he did not pay much attention to it. But he did remark about what he had seen and on July 4 he, his wife, Vernon and a daughter, Betty, age 14, went back to the spot and gathered up quite a bit of the debris.

The next day he first heard about the flying disks, and he wondered if what he had found might be the remnants of one of these.

Monday he came to town to sell some wool and while here he went to see sheriff George Wilcox and "whispered kinda confidential like" that he might have found a flying disk.

Wilcox got in touch with the Roswell Army Air Field and Maj. Jesse A. Marcel and a man in plain clothes accompanied him home, where they picked up the rest of the pieces of the "disk" and went to his home to try to reconstruct it.

According to Brazel they simply could not reconstruct it at all. They tried to make a kite out of it, but could not do that and could not find any way to put it back together so that it could fit.

Then Major Marcel brought it to Roswell and that was the last he heard of it until the story broke that he had found a flying disk.

Brazel said that he did not see it fall from the sky and did not see it before it was torn up, so he did not know the size or shape it might have been, but he thought it might have been about as large as a table top. The balloon which held it up, if that was how it worked, must have been about 12 feet long, he felt, measuring the distance by the size of the room in which he sat. The rubber was smoky gray in color and scattered over an area about 200 yards in diameter.

When the debris was gathered up the tinfoil, paper, tape, and sticks made a bundle about three feet long and 7 or 8 inches thick, while the rubber made a bundle about 18 or 20 inches long and about 8 inches thick. In all, he estimated, the entire lot would have weighed maybe five pounds.

There was no sign of any metal in the area which might have been used for an engine and no sign of any propellers of any kind, although at least one paper fin had been glued onto some of the tinfoil.

There were no words to be found anywhere on the instrument, although there were letters on some of the parts. Considerable scotch tape and some tape with flowers printed upon it had been used in the construction.

No strings or wire were to be found but there were some eyelets in the paper to indicate that some sort of attachment may have been used.

Brazel said that he had previously found two weather observation balloons on the ranch, but that what he found this time did not in any way resemble either of these.

'I am sure that what I found was not any weather observation balloon,' he said. 'But if I find anything else besides a bomb they are going to have a hard time getting me to say anything about it'."

Besides the above press reports in the aftermath of the crash, the Army literally terrorized the local Roswell community, even separating parents from their children and telling them that if they talked about it, they would kill their parents, and vice versa. Nobody spoke about it again until many years later. We will touch upon that subsequently, but it is first important to address the major changes that took place in the wake of the July 1947 crash at Roswell.

CHAPTER THIRTY-THREE

1947

In 1947, the Japanese had forty divisions on mainland China, a condition that began in 1945 and lasted up to 1952. The Chinese civil war between Mao's communists and Chiang Kai-shek's Nationalists was very much in full swing. While the Americans were still very much at war with the Third Reich in exile and while overt hostilities with the Japanese had ceased for the most part, it would still be four years before the *Treaty of San Francisco* would be signed. As the Russians had no nuclear capabilities at that time and their army had been devastated, they were not considered a major threat. On the other hand, the Japanese were the only armed nation on Earth that could attack the United States with nuclear power, and they had also clearly demonstrated their biocidal capabilities by landing their aero-dreadnoughts at Tonopah Army Air Field. The Roswell crash was an embarrassment on many levels, and this included the complete and unwanted destruction of a Japanese super-dirigible.

1947 was a pivotal year in many respects, and it was during this time that the entire military and security structure of the United States was both reorganized and reinvented. So much of this was influenced heavily by, if not a direct result of, what they had learned and acquired from their enemies.

Public relations was always a considerable part of American strategy, particularly through the Office of War Information, but it was taken to a new level with the defeat of the German homeland and the vast array of captured resources which included both documents and key personnel. This included the legacy of Joseph Goebbels, Reich Minister of Propaganda, who was arguably the greatest propaganda maestro in the history of the world. While Goebbels is mostly known for his mass rallies and outrageous films, one of his most effective strategies against the Allies was to suggest and lead his enemy to believe that the Third Reich had been contacted by aliens who were giving them advanced technology. Goebbels, who was in charge of his own military troops and had the rank of General, was responsible for inventing the "Foo Fighters", electrically charged "Feuerballs" (fireballs) that were designed to distract, disrupt, and inject fear into enemy pilots. They were also reported to jam electrical circuits in aircraft so as to cause them to stall and crash.

The Foo Fighters were highly maneuverable, radio-remote-controlled magnesium flares with super-elongated coils to provide extensive duration of the flare's burning. It was the aerial equivalent of a nautical mine, with all the spikes sticking out in every direction in flight. These fireballs were used in conjunction with a cleverly crafted Goebbels claim that, as the Germans were the first government to use television broadcasting of ceremonial events, their signals were intercepted by the extraterrestrials that had been flying in the vicinity of our solar system, and this got them recognized as "the" government of the planet. That is why, at the end of the war, these fireballs suddenly appeared and were buzzing bombers everywhere in Europe. There were also some in the Pacific Theater.

It was through this that the Reich's Propaganda Minister, Dr. Goebbels, was able to force the British and the Americans into standing back. The American records I was looking at claimed that the British were more susceptible to this propaganda because of their nervous strain from years of warfare. The Americans claimed that they themselves were much more skeptical. Personally, I think that the Americans were just trying to push it off onto the British, but they were just as afraid. Nobody had seen Feuerballs before, and they were very afraid that these were actual extraterrestrial craft. It was a propaganda victory of enormous proportions. Dr. Goebbels was able to convince the Allies that aliens were on the Nazis' side and recognized their government alone. The Allies therefore had to allow the entirety of the Third Reich's highest echelon to escape, save for cases of suicide. I can guarantee through the records that I have dealt with that Dr. Paul Joseph Goebbels and Reichsleiter Martin Bormann did indeed escape. It was a propaganda victory of monumentally decisive proportion.

In Germany, Goebbels' umbrella of influence was a complete domination of the broadcasting air waves, the press, the film industry, the arts, and censorship in general. The Americans were to use all of his work as a model upon which to form the infamous Operation Mockingbird.

The Americans had already tested the alien scare technique as early as 1938 with the Orson Welles radio broadcast of *War of the Worlds*, and they knew people were highly suggestible. It was used in 1942 after the Battle of Los Angeles, and FDR even purposely "leaked" the previously shown document to increase the suggestibility of the public, and that included the Congress as well as those of his own people who were not in the know.

As soon as the initial report of the Kenneth Arnold sighting was disseminated in the popular American media, it was read by aviation legend Orville Wright who made a public response. Observing the obvious, Wright's initial statement was that it was all propaganda for the war and to excite the people to get them to believe a foreign power had designs on this nation. As a primary pioneer of American flight, Orville Wright could see that the flying saucer would be used as propaganda to create an entirely new branch of the American military: the U.S. Air Force.

The Roswell Incident followed the Kenneth Arnold sighting, and within three weeks, George Gallup* conducted surveys across the United States, from July 25 to the 30th, and it was discovered that 90% of the population had an awareness of flying saucers. This was the largest awareness ever recorded. This finding told them that their previously initiated public relations tactic with Roswell could be even more effective than they might have otherwise dreamed. The military took advantage of that in 1947 and also many years later.

As was already alluded to earlier in this book, the U.S. Army was completely in charge of the entire Roswell Incident because there was no Department of the Air Force at that time. The flying saucer and alien publicity stunt that was initially used on the public and then recanted almost immediately has left a chilling effect on the popular culture ever since. As much as they learned from Goebbels, the Americans were not exactly amateurs at contouring public opinion. The fact that they bullied the local population and offered such a lame explanation only reinforced the conviction of the population that the original report was indeed the truth. In addition to that, the Roswell Incident was also used as a propaganda coup that the aliens had landed and needed to be fought off with an entire new branch of service that would be borne out of the Army that would be a black budget black hole with no senate oversights. This is quite literally how the U.S. Air Force actually came into being. Even so, the Air Force was only one department that emerged as a direct result of the issues that had arisen in the wake of the Roswell Incident and the numerous circumstances that surrounded it.

Two months after the Roswell Incident, the National Security Act of 1947 went into effect on September 18, 1947. In what was a major restructuring of the United States government's military and intelligence agencies, it created the Central Intelligence Agency, the National Security Council, and the United States Air Force. It also provided an opportunity to facilitate the back channel demand of Emperor Hirohito that the Department of War be abolished and that the United States

---

*George Gallup was the inventor of the Gallup poll, a successful statistical method of survey sampling for measuring public opinion that was highly reputed until he struck out in 1948 by predicting that Thomas Dewey would defeat Harry Truman by between five and fifteen percent. He was employed as a vital component of the OSS during WW II and his services were used as a key part of psychological operations as the OSS was continually trying to assess the morale of various populations, including its enemies and its own people.

would never be allowed to declare war ever again. Accordingly, the Act merged the Department of War (renamed as the Department of the Army), the Department of the Navy and also the U.S. Air Force into the National Military Establishment (NME) which would be headed by the new post of Secretary of Defense. An amendment in 1949 turned the NME into the Department of Defense, a term people are more familiar with.

The United States, which has never actually declared war against a sovereign nation ever since, is bound to the charter of the United Nations not to declare war without the consent of the U.N. Security Council. If they were to do so, they would become an outlaw nation. Without the stipulations in the U.N. charter and the National Security Act itself, the *Treaty of San Francisco* would not have been signed by the Japanese. It took six years after the Nagasaki bombing for all of the different issues to be addressed and solved.

As the military utilized Project Paperclip to integrate the German rocket scientists into America's fledgling rocket program, the formation of the CIA enabled the integration of the German intelligence model provided by Reinhard Gehlen, a German general who was repatriated to serve as the primary consultant in the formation of the CIA. This also paved the way for programs such as MK-ULTRA and Operation Mockingbird, the latter using the techniques of Goebbels to embed Hollywood and other media to serve the agenda of the agency, using them as instruments of psychological warfare. It was from this base that a slew of flying saucers and alien movies were unleashed upon the public, most of them in the early 1950s. Although most of these were B movies, they embedded the idea of alien visitors deep into the public's psyche, very effectively reinforcing the propaganda that had accompanied the Battle of Los Angeles and the Roswell Incident. It is a lasting imprint that remains impressed upon our culture unto this day.

After the various seeds about aliens invading the Earth and a flying saucer having crashed at Roswell had been effectively planted in the minds of the population, things pretty much calmed down with regard to the explosion of the Japanese super-dirigible in 1947. In other words, it was not much of a public relations problem as Project Bluebook was defraying any assertions about possible UFOs. All of this might have remained a non-issue if it were not for the Apollo moon landing in 1969.

CHAPTER THIRTY-FOUR

BARGAIN WITH THE DEVIL

When Apollo 11's crew landed on the moon in 1969, it was broadcast on both national and international television. It was a very exciting time for Mankind. One of the millions of people who shared in the excitement was Sergeant Melvin Brown, an army cook who had participated in the initial recovery of the aircraft that crashed near Roswell.

Watching the first moon landing with his family, Brown became very excited and started talking to the point where he could not control himself. He told his family that he was once in a truck where there were enemy aliens under a tarp. Although he was not supposed to, he looked under the tarp and saw the bodies. They were yellow/orange but had turned gray with rigor mortis. He recognized them as being Oriental, but he said that a fireman had described them as clearly being Japanese. His superiors, however, had instructed him to say they were aliens. This, however, was retracted the next day. All of this excited his family, and it generated a lot of excitement as well as rumors. Brown had to warn his family that these matters must not be mentioned any further or it could cost him his very life.

With some of the cat out of the bag, the Army had to go into damage control mode. Old wounds were being exposed and stirred up, and the military high command decided they had to get somebody to spin this story so people would not find out they were guilty of war crimes, not to mention the degree to which they had covered up the lies concerning American's concessions in World War II. Accordingly, William Blanchard and his team involved with Roswell called on one of the Army's top specialists in PsyOps (Psychological Operations) and brought him back from Vietnam where he had demonstrated remarkable success in this capacity. His name was Michael Aquino, an avowed satanist who broke off from Anton LaVey and the Church of Satan and founded the Temple of Set. Although he was never convicted, Aquino would later forever be associated with the child sexual abuse scandals of the 1980s at the Presidio in San Francisco. Aquino himself was a "social engineering" expert on warping minds and controlling populations.

As a Department of Defense Research Librarian at the Presidio Military Base of San Francisco, one of my duties was to service the military officers who frequented it. One of the library's most frequent patrons was Lieutenant Colonel Michael Aquino, a key military intelligence officer. He would often request ancient magical manuscripts or "grimoires" from foreign libraries. As these materials were too valuable to leave the library, they had to be perused on site. As Aquino's purpose was to conduct rituals with his coven, the only place he could do this without disturbing other patrons was the library itself, after closing hours. This is the same place where I was assigned to burn classified documents in the incinerator. While I carried out my job, I was also being exposed to occult rituals and the accompanying energies associated with such.

Very impressed with my ability to correlate information and to retrieve the materials he needed, Aquino became very friendly with me, seeking me out to be his disciple and also sharing a considerable amount of his knowledge. He liked to show off. While there is considerably more that can be said about my interactions with him, they will be the subject of a subsequent book. What is important in this narrative is that I share his substantial role with the Roswell cover-up. It is of note that, despite his salacious and diabolical reputation, he was promoted to be Deputy Director of the National Security Agency, the very organization that, amongst others, was formed as a result of the Roswell Incident.

STANTON FRIEDMAN WITH HORNED EYEBROWS

BARGAIN WITH THE DEVIL

In order to effectively distract the public, Aquino tapped a fellow cultist from the local community who was ready, willing and able to rise to the call and reinforce the myth that the Army had long ago created about the Roswell Crash. This was Stanton Friedman, a fellow member of the "Cult of the Horned Brow" who made his mark in the UFO field and especially with regard to Roswell, and so much so that he was often referred to as the "Father of Roswell". As he was such a major player in the Roswell cover-up, it is important that I divulge some of his background.

In the 1960s, Friedman was working for General Electric's Aircraft Nuclear Propulsion Department which had a publicized $100 million Air Force contract employing 3,400 people (including 1,100 engineers and scientists) to develop nuclear aircraft. In this capacity, he worked on a black budget scam in what was so-called nuclear flight. They converted a Convair NVAH-36 bomber, a "bucket of bolts" that was purposely fitted with a leaky nuclear reactor that they flew all over the United States. The pilots they used were shielded from the reactor with very thick lead, and they recalled very old pilots as younger airmen would become sterile as a result of the leaked radiation. This program irradiated so much of the American landscape with its low level flight paths that Stanton Friedman was literally frightened to the point of leaving the country, heavily motivated by reason of the fact that he sired a developmentally disabled child by reason of his own exposure to this radiation. Friedman subsequently got himself Canadian citizenship where the child could be maintained without cost in their free welfare medical system. That is how bad it was.

Although Friedman had loyally served him in spreading the gospel of Roswell, Michael Aquino was amused to no end by the horribly ironic fate of the afflicted child and the tragedy that befell the entire family. He even laughed at the karma of it all, attributing it to Friedman's role in irradiating so much of America and thereby producing so many children like his own. Aquino thought that he himself, as a master of satanism, was well above the wheel of karma.

The actual amount of how much was sunk into this diabolical nuclear project went well beyond the hundred million figure, even reaching into the billions. This is an example of the kind of black budget scam that the military was able to get away with through the Air Force. It is no coincidence that Aquino would eventually find his way into becoming a high ranking member of the National Security Agency, another department created by the National Security Act of 1947.

For years, Stanton Friedman served as Aquino's spot man and took to the task at hand as a real pro with regard to steadfastly and diligently working to create a smokescreen of disinformation about Roswell. A cunning perpetrator in creating the Roswell mystique, Friedman became a famous author on UFOs who was routinely glad-handed by major media and was a regular speaker on what became the very worn-out lecture circuits.

In addition to Friedman's tireless efforts, two authors with intelligence backgrounds breathed new life into the Roswell story with the 1980 release of *The Roswell Incident* by Charles Berlitz and William Moore. There was new information being circulated about the crash with individuals coming forward for interviews, turning Roswell into a rallying cry for UFOlogists and true believers. For many, the Roswell Incident has served as the Holy Grail of UFOlogy. Many people believe in it just like a patriot believes in the American flag. It is, without a doubt, the most famous of all UFO stories, but the myth would grow to even larger proportions. This has to do with rather ordinary political history but one that is overlooked and undervalued by the American public.

Aquino made a point of telling me that when he was called back from Vietnam in 1969, Roswell was suffering from the fact that, when Lyndon Baines Johnson was still in the running to be re-elected as President for the 1968 election, he purposely arranged for the Roswell Army Airfield and other military bases to be moved to Texas. The reason for this was to garnish more votes for himself in the state of Texas. This is how politicians connected to Texas seek to wield influence so as to become President. In terms of the Roswell Army Air Field being abandoned, the town was going to wither up and die because the entire Roswell economy was based upon the military. While still suffering from this situation, Michael Aquino approached them and offered the Roswell residents a deal with the devil. He told them that if they would accept the story he would tell, he would make sure they had all the financing they would ever need and would never need a military base again. They were going to do for Roswell what Medjugorje did for Bosnia, making it the shrine of what amounted to a new religion based upon a legendary UFO crash, and that is exactly what they turned Roswell into. Together, they said it was an alien phenomenon and they got everybody to visit there. This Roswell myth floats their entire educational system, floats their entire town, and they are more or less willing to kill or die to protect this myth and keep their economy going because it has given them a quality of life they never could have dreamed of and all without even having a military base on site. This is the kind of cynicism and evil that we are speaking of. It is also something that I have plenty of proof for because so much of my life has been deeply affected by all of this.

The International UFO Museum and Research Center was opened at Roswell in September of 1991 with none other than Stanton Friedman as the ribbon cutter. Aqunio's primary ally and facilitator in Roswell was the founder of this museum, Glenn Dennis, a mortician trained at the San Francisco College of Mortuary Science who was also the first witness to confirm claims of alien bodies at the Roswell base itself. While his claims were featured prominently in Friedman's book, *Crash at Corona*, his credibility was entirely discounted after serious pro-UFO researchers found far too many inconsistencies in his stories.

These included the false identity of a nurse who was said to be a witness to the alien bodies. When the name he gave was proven to be phony, he said that was because he had promised that he would never reveal it to anyone. If that were the case, he certainly did not have to use any name at all. By the fact that he consistently and knowingly provided false information, he could not be taken seriously. Mind you, this is the founder of the museum. It is interesting to note, however, that in a book, *Witness to Roswell: Unmasking the 60-Year Cover-Up* by Thomas Carey and Donald Schmitt, the authors, who themselves unmasked Dennis's false assertions, did note that they had found credible witnesses that claimed Dennis had told them about the Army asking him for child-sized caskets way back when it happened. This is not only consistent with the story line I have offered you, it demonstrates that he was clearly in the pocket of Michael Aquino as described above.

Dennis's efforts were significantly augmented by Roswell Mayor Thomas Jennings, a marketing major with an entrepreneurial spirit so formidable that commercial exploitation of the UFO crash was featured in *Forbes* magazine. An official statement from Jennings' office, as quoted in the book *The Roswell UFO Crash: What They Don't Want You to Know* by Kal K. Korff was, "the Roswell UFO myth pumps an additional five million dollars a year into our local economy. That's why we embrace it." *Forbes* magazine's article of July 15, 1996 by William P. Barrett stated that Roswell's civic leaders were positively giddy about the prospects for the upcoming 50th anniversary of the crash. Forbes also pointed out that the small city had a history of hustling as its founder was a professional gambler who named it after his father, Roswell.

Like Hollywood, Roswell is in the fantasy business. Top military officials in 1947 disavowed the captured saucer story within hours, saying that what landed was the debris of a weather balloon. The Pentagon now says the balloon was connected with monitoring Soviet nuclear testing, but the Soviets could not possibly have conducted any such testing until 1949. By the mid-1980s, people began to recall seeing dead or dying aliens in varying dispositions. These accounts have since been embellished.

With the Roswell museum operations in full swing in the early 1990s, the rampant rumors that the Government had recovered alien bodies and was engaged in what might be considered the greatest conspiracy cover-up of all time had gained a tremendous foothold in popular culture. There was so much public consternation over the issue that the Government itself took the very remarkable and unusual step of producing two reports on the subject in an attempt to close the case and once and for all put the matter to rest.

Assembling and declassifying thousands upon thousands of documents relating to the Roswell incident, the Air Force released *The Roswell Report: Fact vs. Fiction in the New Mexico Desert* in 1994, stating that it would "tell the Congress and the American people everything the Air Force knew about the Roswell claims." The report itself was nearly one thousand pages long.

In 1997, just before the 50th anniversary of the Roswell Incident, the Government released a second report entitled, *The Roswell Report: Case Closed.* This muddled the matter even further because it stated that eyewitness accounts tied to the 1947 recovery had actually occurred years later, thus confusing and further strengthening the Roswell Incident's hold on the public's imagination.

> "Air Force activities which occurred over a period of many years have been consolidated and are now represented to have occurred in two or three days in July 1947," the report said. "'Aliens' observed in the New Mexico desert were actually anthropomorphic test dummies that were carried aloft by U.S. Air Force high-altitude balloons for scientific research."

None of this did anything to dissuade the public. In fact, the reports sounded so ridiculous that it appeared the Government was doing everything it could to hide the "truth" of alien contact. The Air Force did not even exist at the time of the Roswell Incident. It was a perfect psychological operation because the public was believing what they wanted them to believe in the first place.

Despite the military's assertion that the Roswell Incident was a side effect of Cold War secrecy and sci-fi fantasies, the story retains a vital spot in UFO lore. The town of Roswell has turned into a tourist destination, hosting the International UFO Museum and Research Center and an annual Roswell UFO Festival.

More fuel was added to the fire of the false public relations scam that is Roswell when an army intelligence officer named Lieutenant Colonel Phillip James Corso came out of the woodwork to publish *The Day After Roswell* with Bill Birnes as his co-author. Having worked at the Pentagon with security clearance, Corso was at this point in his life suffering from senile dementia, and claimed that he was visited by space brothers. He approached Bill Birnes, who eventually owned *UFO Magazine*, and told him the space brothers had been visiting him and told him that we could have world peace. This, however, was not a very sellable angle for a book. Instead, they approached Senator Strom Thurmond.

Told that Corso wanted to write a book entitled *I Walked With Giants* about how Corso had served Eisenhower and the Supreme Allied Command in Europe during World War II, Strom Thurmond was asked if he would be kind enough to write an introduction for the book. Senator Thurmond was amenable to the request and wrote a flattering introduction about Corso. Birnes, however, stuck that in the Roswell book. As soon as Senator Thurmond found out, however, he was bewildered and angry and threatened to sue them both if they did not immediately withdraw all copies that had been printed of *The Day After Roswell*. Somehow, this was able to be done within a few days. It is amazing that this could be accomplished in time for the 50th anniversary of the Roswell incident. With government connections, the book was withdrawn, incinerated, and completely replaced within 48-72 hours of the threat and placed onto the shelves of Barnes and Nobles, Crown, Super Books and all the rest; this time without Senator Strom Thurmond's introduction. How can this be done so fast without the Government's cooperation? This shows how willing the Government is to push this crap down your throat. Accordingly, people blindly believe that technological products such as pacemakers and microwaves are a result of the reverse engineering of alien technology. It is only alien technology in the sense that it comes from the alien or foreign civilization that is indeed Japan.

Corso died with a year of the publication of his book and left his family subject to scandal and lawsuits.

THE ROSWELL DECEPTION

CHAPTER THIRTY-FIVE

INTERNATIONAL RECOGNITION

In the 1990s, the Hakui UFO Museum was built in Japan. Hakui is located on the Sea of Japan right across from Korea, and over fifty million dollars has been sunk into what was once billed as the only UFO museum in the world. While many of the exhibits are corny beyond belief, such as a model of an extraterrestrial "alien" from Roswell, it also features exhibits from NASA and other space programs.

As soon as the museum announced plans to open, the FBI took a very keen interest because the museum had an open door policy in seeking to attract new information and exhibits. This had the potential to drive a very big wedge into the secrecy that had so long been held from the American public with regard to World War II and the Roswell Incident. Press reports, however, suggested that the FBI's interests were focused on a memo sent to J. Edgar Hoover by Special Agent Guy Hottel from the FBI Washington Field Office on March 22, 1950.

> "An investigator for the Air Forces states that three so-called flying saucers had been recovered in New Mexico. They were described as being circular in shape with raised centers, approximately 50 feet in diameter. Each one was occupied by three bodies of human shape but only 3 feet tall, dressed in metallic cloth of a very fine texture. Each body was bandaged in a…
>
> "According to Mr. [name blackened out] informant, the saucers were found in New Mexico due to the fact that the Government has a very high-powered radar set-up in that area and it is believed the radar interferes with the controlling mechanism of the saucers.
>
> "No further evaluation was attempted by SA [blackened out] concerning the above."

Not surprisingly, this document is, according to the FBI, "the most viewed document in the FBI Vault, our online repository of public records." This could be taken as demonstrative of leaking a spurious document to a gullible rubber-necking public in order to distract them from any suspicions of what I have shared in this book, but it is more accurately to be taken as exemplary of the interdepartmental miscommunications reflecting "distorted echoes" that reverberate years after the fact.

Perhaps nothing would have come of any of this and the FBI could have laid its concerns to rest if it were not for the fact that, as the 50th anniversary of the Roswell crash in 1997 was approaching, the museum announced their plans to open a Roswell exhibit to honor all of the Japanese prisoners who had died at Roswell. When I learned of this, I anonymously sent them pieces of the Roswell wreckage which I had absconded with from my days at the Presidio. It was nothing more than Japanese aluminum, but when they announced it, it was to be featured as the center piece of the Roswell exhibit in Japan.

To remind you of what I said earlier in the book, the debris I had absconded with was the result of me being forced to work with the Roswell files, all of which were above my security clearance and pay grade. The Presidio base commandant had ordered my superior to present him with certain files from the Roswell boxes. As it would takes weeks and weeks to do this job, my lazy Operations Manager, Andrew Minjiras, did not want to do it as it was too much work.* Accordingly, he bullied me into doing it by stating that if I did not do it, he would accuse me of attempting to walk out with the materials and thereby have me imprisoned for the rest of my life.

The file boxes said "crypto cosmic top secret" and over those words was written "above", meaning that it was literally above "crypto cosmic top secret". While such a designation sounds very "alienesque", everything in the files was quite terrestrial. With the lackadaisical attitude about security exhibited by my Operations Manager, however, and when the opportunity presented itself, I took action. Sorely vexed to point of treason, I took it upon myself to abscond with the most coherent fragments of the Roswell debris that the boxes contained. A sense of righteousness motivated me to send the salvage to the people of Japan to serve as the centerpiece for the Roswell exhibit. I covertly consigned the debris through Teleport USA, a banking and promotional organization which facilitated business between Hakui City and Los Angeles.

---

*The Presidio Postal Library Staff that I worked with were as follows: Juanita Taylor was the overall director of the Department of Defense Library. From the deep South, she was functionally illiterate, having obtained her post as a result of nothing other than affirmative action, thereby possessing the highest security clearance under which documents destruction was done. Her "English" was a distinctively regional Creole and impossible for anyone to comprehend outside of those of us who invested the time and effort to do so. Besides Minjiras, there was my immediate superior, Carolyn Garrett, an African American woman who assertively differentiated herself from Juanita Taylor by emphasizing her education and professional competence. There was also Frank Conway, an assistant librarian who was always quite drunk on duty but nonetheless was an individual whom I found to be most sympathetic.

HISTORICAL MARKER COMMEMORATING ROSWELL P.O.W. CAMP

An incredible amount of excitement was generated as a result of my entrusting this debris with the museum. Although no one in the general public really knew the details of what was involved, rumors were circulating on the UFO circuit like wildfire. It was a real BIG deal and was being billed as the UFO event of the century. Everyone in UFOlogy embraced all the excitement.

The late Robert Dean, a famous UFO investigator, divulged the news on Ken Schram's popular Seattle television show "Town Meeting" and British crop circle guru Colin Andrews announced throughout the Commonwealth that the opening of the Hakui UFO Museum was going to feature alien UFO debris. None of these UFO oriented fans or researchers realized, however, that the intent of the Japanese was to expose American war crimes and memorialize their victims.

Teleport USA took the already extant fever pitch to a new level by promoting a tourist package which included a trip to Hakui and the UFO museum as well as a convention with prominent UFO speakers. Considerable fuel was added to the fire when Art Bell, host of the popular radio program "Coast to Coast A.M.", got into the act.

On April 18 of 1996, Art Bell received what became known as "Arts Parts". Totally identifiable as being of the same materials I had sent via Teleport, the pieces they received arrived with a long letter from a person who claimed to be serving in the military with a security clearance and did not wish to go public and thus risk losing his career and commission. He said that his grandfather was involved in the Roswell recovery and kept some of the parts which he identified as "pure extract aluminum". As his grandfather's estate was now settled, he was in possession of the parts since 1974, and "after considerable thought and reflection", he was giving them to Art to share with those in the UFO research community.

Linda Moulton Howe utilized energy-dispersive X-ray spectroscopy in order to test the material and determine what its chemical components were. Such a test is very important because everywhere in the universe we are all made of the same star stuff, but like a fingerprint, different star systems have different gradations of the material that makes up our solar systems. That is how we can tell if a different meteorite comes from a different solar system. With energy-dispersive X-ray spectroscopy, we can tell whether something is of this Earth or not. The testing was done quite publicly, and on May 19, 1996, Howe shared the results on the air. She announced, to everyone's surprise but my own, that "Art's Parts" were nothing more than Japanese aluminum. It was now proven on air that the type of debris Art Bell was given were all quite terrestrial, and it was exactly the same as what I had shared with the museum. No one, of course, had a clue as to what Japanese aluminum truly represented.

The revelation that it was Japanese aluminum burst the bubble of the numerous fans of Art Bell who had sucked all of this up. On the other hand, there were others who openly and publicly ridiculed the ongoing drama that Bell was perpetrating and the assertions of this anonymous individual who had sent the material in the first place. Although the bulk of Bell's audience was deflated, the anonymous provocateur remained as vigilant as ever, paying close attention to the audience response.

After hearing the results of the tests, the anonymous "military man" who had started all of this nonsense offered a follow-up letter stating that the "pure extract aluminum" he had sent them was used as a conductor for the electromagnetic fields created in the propulsion systems. If this was not enough, he also stated that critically-needed data was "eliminated" by the self-destruct mechanisms on the disc vehicle itself; and, "Furthermore, the occupant-survivor of the crash, refused to disclose technical information, despite a series of interrogative attempts to extract technological data. No means could be found to secure the information."

He further explained that the reason they could interrogate the survivor, who was found in the crash itself, was that he could communicate to humans telepathically.

INTERNATIONAL RECOGNITION

Further assertions in the letter were that, according to his granddad's notations, the self-destruct mechanism "was to insure that in the event of crash or capture, that no verification as to the alien network or the home world confederation could be proven, and a compromise of the Alien directives transpired. Apparently, 'the probe ships were constructed with metallic base metals, indistinguishable from Terran metals, as a protective measure & security safeguard'."

It was now clear that Art Bell had led his audience on a wild goose chase. To the man who had sent it, who was none other than (then) Lt. Colonel Michael Aquino himself, it was a big joke. Here he was, giving them the whole secret (the fact that it was Japanese aluminum from the dirigible), but they were too ignorant and stupid to understand it. This is how Aquino's diabolical sense of humor worked. He was not only deflating the public's excitement and making fools out of them, he was also making a public fool out of Art Bell, a former military man who directly followed Aquino's orders with regard to who could and could not be on his show. Aquino relished making fools out of his own lackeys. This is the mind of a true devil.

There was another aspect to Aquino's psy-op, none of which registered on Japan's radar nor in any way impacted their response to U.S. provocations at the international level. He was engaging in preemptive domestic damage control in order to buffer what impact might have imposed itself upon the American psyche had the Japanese followed through with opening their museum under its originally intended theme of serving as a memorial for their prisoners of war.

Irrespective of Aquino's fun and games, the news of "Art's Parts" being exposed reverberated around the world, and there was a serious and ominous response that carried considerable backlash. The FBI dispatched a team to Japan in order to recover the aluminum that I had sent. The metal was stolen federal property and they wanted it back. This is how seriously the Government takes the issue of preventing the truth of the Roswell Incident from ever being exposed.

Serious contention ensued between the Japanese and American governments. Miscommunications exacerbated the issue, and a lot of it had to do with language. The Japanese were intent on building a memorial museum for their dead prisoners of war who had suffered at Roswell.

After the Japanese refused to give the material back, President Clinton, in order to protect all of these old war criminals in the American High Command, told the Japanese that he was going to ratify a trade agreement. This was in 1996, and he did indeed do so. It was the toughest trade agreement that had ever been imposed upon Japan since the *Treaty of San Francisco* was signed on September 8th of 1951 at the Presidio. Clinton's action ended many of Japan's import quotas against American goods, all of which had originated in Japan's Pacific War victory condition.

Japan exacted its revenge immediately by ratcheting up its divestment from the United States and beginning to invest substantially in China. This directly caused the Chinese economic boom because they were creating industrial systems for the Chinese that they simply just did not have. Japan had all the know-how and materials for setting up production lines, assembly lines and whatnot. In other words, they invested in China by teaching them how to manufacture and supplying them with the necessary infrastructure.* This divestment from the United Stated resulted in a depressed economy in the U.S. for many years. In the meantime, China has boomed ever since.

Through the simple act of sending the Japanese museum one of the most highly emotionally charged pieces of scrap metal ever known, I ended up having an impact on the entire world.

Over the potential vulnerability of having their dirty secrets exposed, the United States had sent a strong message to Japan. Their response, however, did not involve saber rattling. The Japanese simply responded economically by quietly undermining the U.S. economy. In this respect, the United States was far more concerned about keeping their precious secrets than they were about the economic welfare of their own citizens.

All of the excitement over "Art's Parts" and the Teleport event, however, had exhausted the patience of the radio audience. People

---

*The reason you do not hear much about Japanese manufacturers these days is that the best of them have moved from manufacturing consumer goods to concentrate on so-called "producer's goods", items that, though invisible to the consumer, happen to be critical to the world economy. Such goods include the highly miniaturized components, advanced materials, and super-precise machines that less sophisticated nations such as China need(ed) to produce final consumer goods. The label on everything from cellphones to laptop computers may say: "Made in China", but they are actually dependent upon highly capital-intensive and expertise-intensive manufacturers in Japan who have quietly done much of the most technologically demanding work in order to produce the producer's goods, without which, the final consumer goods could not exist. For further information, refer to the book *Mechatronics: Japan's Newest Threat* by V. Daniel Hunt (Advanced Industrial Technology Series). "Mechatronics" is a term coined by the Japanese to describe the integration of mechanical and electronic engineering. The concept may seem to be anything but new as we can all look around us and see a myriad of products that utilize both mechanical and electronic disciplines. Mechatronics, however, specifically refers to a multidisciplined integrated approach to product and manufacturing system design. It represents the next generation of machines, robots, and smart mechanisms necessary for carrying out work in a variety of environments – primarily, factory automation, office automation, and home automation.

could take no more of it, and it was a huge disappointment for everyone involved. Shortly after Howe announced the test results, Teleport USA made the following announcement on June 12, 1996:

> To whom it may concern:
>
> Re: Cosmo Isle Hakui Museum
>
> We regret to inform you that the Cosmo Isle Hakui Museum Symposium scheduled for July 19-21 and the tour that was set for July 16-23 has been cancelled. At this late stage in time, it is very unfortunate that for political and bureaucratic reasons this symposium has been canceled.
> There is opposition and controversy regarding the usage of the word UFO and some of the material that was set to be displayed. Museum officials have decided that now is not the appropriate time to hold this symposium. Museum officials are looking to reschedule the symposium at a later date.
> Since no official statement has been made about the Grand Opening, it is anticipated that [the museum] will open July 1. Please be aware that the Opening Ceremony is by invitation only.
>
> Thank you for your understanding.
>
> Sincerely,
>
> Teleport USA

This resulted in all sorts of conspiracy theories being generated, most speculating that the United States Government had pressured Japan to back off, especially in regards to the metallic debris from the "flying saucer" crash. This tact, of course, suited the Americans perfectly. It sounded to most people to be a UFO cover-up rather than the hiding of the Japanese dirigibles that had been operated by prisoners of war.

It was not until fifteen years after the incident that I was in a position to come out as a public informant, putting the aluminum wreckage I still possessed on display so people could see it. I had them put on gloves so that oil secretion would not damage or compromise the paint or symbols. This presentation was on January 29, 2011 and was recorded and put out in DVD format as *Roswell and the Rising Sun*.

Everything was now out of the bag, and although it was fifteen years after the museum fiasco, I received a nasty blowback. Coming home from the presentation, I came home to find that my mom had her aorta dissected in two places with the precision of a laser scalpel and no incriminating marks. Prior to that point in her life, she was already functionally incapacitated. She was now good as dead, and although the aorta should have ruptured, it did not. I received a lot of public support as this was made known and she lasted until March 19th, the same day Annie Jacobson's book on Area 51 was released. In its own specific way, her book was a response to what I had just released.

Jacobson's book, *Area 51: An Uncensored History of America's Top Secret Military Base*, is a book that was hailed by critics, including the *New York Times*, and is described by *Slate* (magazine) as "A compelling narrative of 50 years of covert operations by the CIA, the U.S. military, and the mysterious 'Atomic Energy Commission'....Her meticulous research makes for a fascinating read, as it intersperses the accounts of secret government projects with anecdotes from the people who made those projects happen."

Area 51 includes the theme of the CIA taking over the Tonopah Army Air Field so that they would not have to turn over any of the debris or files to the FBI. That was the kind of politics that was going on in the U.S. at that time. Conducting extensive interviews with historical figures who worked at the facility, Jacobson spent at least five years writing the book and did a very thorough job that included abundant sources and footnotes. While the mainstream critics did not slam her, the reviews are very positive on Amazon as well, save for her take on the Roswell Incident.

As Annie Jacobson's book was coming out in the same year as the 50th Anniversary of the Roswell Incident, she was approached to add seven or more pages which were in complete contrast to everything else she had already written about Area 51. Whereas most of her book was exclusively tight journalism, the extra pages included a very bizarre story that was presented as fact, as if this crack journalist had finally discovered the "Deep Throat" of Roswell. Exclusively based on an interview with an anonymous engineer from EG&G (Edgerton, Germeshausen, and Grier, Inc., a United States national defense contractor), Jacobson asserted that the Roswell crash was the result of Dr. Josef Mengele working with Joseph Stalin to create deformed children to fly a Horton Flying Wing and create a "War of the Worlds" panic. Her tale was so confusing and offensive that neither UFO researchers nor debunkers could take it or her seriously. Here was a woman with a stellar degree in reporting and a heretofore irreproachable reputation who had sold out, thus becoming a laughingstock as far as serious journalism is concerned.

Area 51 is now operated in tandem by the CIA and Air Force, and one of their biggest secrets is that they are hiding the history of the prisoners of war who suffered torture and death at Tonopah Army Air Field, Roswell and many other places throughout the American Southwest. There were other crashes in the Southwest where Japanese bodies were recovered, including Ely, Nevada and Kingman, Arizona, the latter featuring "aliens" who were speaking English, as all prisoners of war were forced to adapt to while in captivity.

As alluded to already, the Roswell Incident has served as a shrine and holy grail, drawing curiosity, intrigue and interest worldwide. It has also served as a substantial cottage industry for the city of Roswell, New Mexico. Various serious UFO researchers have picked apart many of the illogical aspects concerning the popular accounts. None of them, however, had my own unique perspective by reason of the thousands

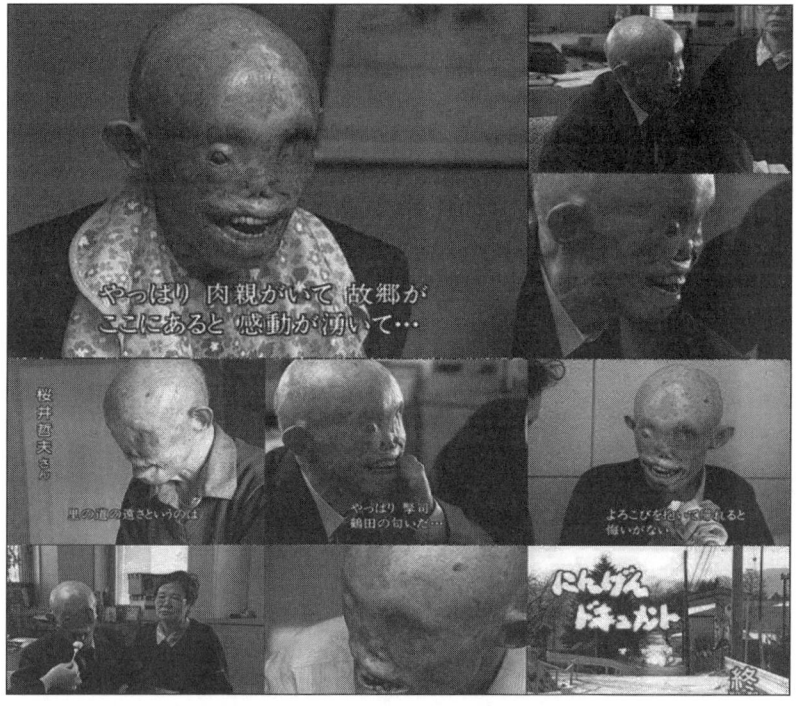

ABOVE IS A JAPANESE PRISONER OF WAR WHO WAS TORTURED BY THE AMERICANS. HIS EXTENSIVE BURNS SUGGEST HE WAS AN ACTUAL SURVIVOR FROM THE ROSWELL CRASH.

upon thousands of documents that I reviewed and destroyed while working as a Defense Department Research Librarian at the Presidio.

In the background of all of this debacle of nonsense, misdirection and psy-ops was (then) Lieutenant Colonel Michael Aquino, later to be the Deputy Director of the National Security Agency. While the National Security Council was put on high alert and the FBI was put in a panic to "patriotically" protect the Government's great secret, Aquino was taunting both sides by exposing the actual wreckage so as to make fools of his own supporters and at the same time deflate the emotions of true believers in UFOs. As he considered himself above the wheel of karma, he was gloating in his own self-created perception of himself. It cannot be denied, however, that he was clever enough to pull it all off. His accomplishments pale, however, in comparison to Emperor Hirohito, a man who overcame tremendous odds and opposition to preserve his own country and put it on top of the world economically. Hirohito destroyed people, but he did so to preserve the integrity of his empire and enhance the life of his people. He was forced to make many brutal decisions, and while he was not so arrogant to believe himself to be divine or above the wheel of karma, Hirohito outlived his contemporary rivals by decades and survived to a ripe old age

CHAPTER THIRTY-SIX

Conclusion

At the beginning of this book, I presented an opening statement that the Government has orchestrated a series of lies around Japan's role in the war and how these played into deceiving the public with regard to the Roswell Incident. If you go back and reread the opening statement, I would expect you to realize that all of the bases have been covered. As far as I am concerned, the case can be put to rest.

In addition to what I have presented in this work, there are considerable supplemental sources you can read. Since 2011, I have tirelessly produced volumes of lectures on the internet covering all of these topics in even more detail. This book is the first time they have been consolidated in a summarized format for ready consumption. Many of my lectures, particularly those I have done for other hosts, are still available online. My own series of successive websites have all been viciously attacked and taken down, and the majority of what I have contributed in terms of lectures is not so readily available any more.

I am well aware that there are certain obdurate people who will make irrational demands that proof is not evident unless they see a "signed surrender" from Harry Truman himself as well as actual photographs of the super-dirigibles. As I was obligated to destroy such, circumstances have not allowed me to present such. Enough circumstantial evidence, however, has been provided herein to conclude the obvious. This includes the words of Secretary of State James Francis Byrnes who, speaking on behalf of the President and the United States, said in his August 11th Dispatch that, "…the ultimate form of government of Japan shall…be established by the freely expressed will of the Japanese people." There is also Truman's admission that hostilities did not end with the ceremony aboard the *Missouri*; and, of course, the fact that the *Treaty of San Francisco* was not signed until 1951.

I will also note that there are certain dishonorable researchers who have exploited materials derivative of the bitter fruits of my own sufferings but have never given myself credit. Indeed; the worst part of all of this is that I have not only received no credit but that they have instead misrepresented what I have put forth and have distorted the information by presenting it in an out of context fashion.

One instance included a fairly well known author who, when confronted about the fact that he was using my material, admitted that he had, in fact, plagiarized it. Asked why he had excluded my name, his answer was that I was anti-American. Before I end, I would like to clear up this matter once and for all. Although I was born in Taiwan, I have given up on my Taiwanese citizenship and am now a full American citizen. I have served in the armed services and have also worked for the Department of Defense. It is my contention that the various people you have read about in this book and their nefarious deeds are the ones who are indeed anti-American. This begins with the communist Jack London, the man who advocated genocide against the Sinitic race and lit a flame under Hirohito in the first place. You have read about the others herein and they are clearly identifiable by their actions and actual history. The fact that many of these nefarious individuals who have attempted to undermine the founding principles of the United States are icons that have been lauded, promoted and embellished by the Office of War Information and its successors is a problem with people's misperception of the truth. It is my sincere hope that the publishing of this book will be a substantial move forward in correcting these misperceptions.

I will further add that what I have to offer is not meant to offend any servicemen who ever fought in war or otherwise served their country. Virtually all of the war veterans who have engaged me after hearing my stories have appreciated such because it gives them a more complete explanation of the reason they suffered such horrors and serves as a sense of closure. A particular case in point is Wouter Hobe, a former prisoner of war. See his e-mails on the following page and also refer to Appendix C. On page 286, you will see the statue dedicated to the young men who were imprisoned in the Bataan Prison Camp as is referenced in Wouter Hobe's e-mail of February 2, 2012.

Lastly, I would also like to state that this book is not intended to be a referendum on the subject of aliens, a subject which has become established folklore throughout the world but is most prevalent in America. This is a subject which is very complex. This particular book is intended to demonstrate that the subject of aliens was used to deceive and mislead the public with regard to the horrible atrocities and lies that have occurred surrounding World War II.

The subject of aliens brings us back to how I ended this book, demonstrating that so much of what transpired around the Roswell mystique was the handiwork of Michael Aquino. While he is most certainly not the be-all and end-all with regard to what is known as alien phenomena, he has played a substantial role in abusing this theme towards deceiving people in programs that extend well beyond Roswell. Accordingly, he will be the subject of my next book which will be based upon the

CONCLUSION

Douglas Dietrich <douglasduanedietrich@gmail.com>　　　Tue, Feb 2 at 5:39 PM
To: Sky Books

---------- Forwarded message ----------
From: **wouter hobe** <wouterhb@gmail.com>
Date: Sat, Nov 19, 2011 at 10:44 AM
Subject: Hallo
To: Douglas Dietrich <douglasduanedietrich@gmail.com>

Thanks for your answer. I was once in SanFrancisco where I met with Roger Mansell. He encouraged me to write down my experiences. After all he was a journalist with lots of experience. Wich I did verbatim the way it came out of my memory. Presently it is being edited in a proper readable format. It is amazing what is hidden in the human mind, that bit by bit will release memories. Presently I am 78 years old. Most of my friends ex POW's are older and vast dying off. To that end it will not be to your benefit to meet any of my buddies. We have "laid our eggs" so to speak and mainly come together to have a good Indonesian meal.
Mind you, the next time we meet I certainly have something to report to them, thanks to you.
Attached you will find a picture of a statue dedicated to the boys in the prison camps. It is in Holland.
It is exactly the way I looked in 1945, tropical sores on the legs and thin. When I exhaled and pulled my tommy in, I could feel my spine. I was wearing a tjawet, a loincloth. We had to run to the bathroom as we had chronic diarrhee, so this loin cloth came in very handy. Because I had no fat left in my body, a part of my colon came out (1 foot) while doing my business which I then had to pull up after.
Us boys had to smuggle meat into the prison camp for the people suffering from beri-beri. On transport out of the camp we had a tube up our anus with money and met a local guy selling meat on our bathroom stints and the meat was then put in a little bag hanging in between our legs. We were never inspected on the crotch upon our return. A little dice of meat would deminish the swelling on these patients. You can now understand that we in North America are overeating with our plate full of steak . Please check out our webside: Kumpulana.ca.
Regards Wouter Hobe.

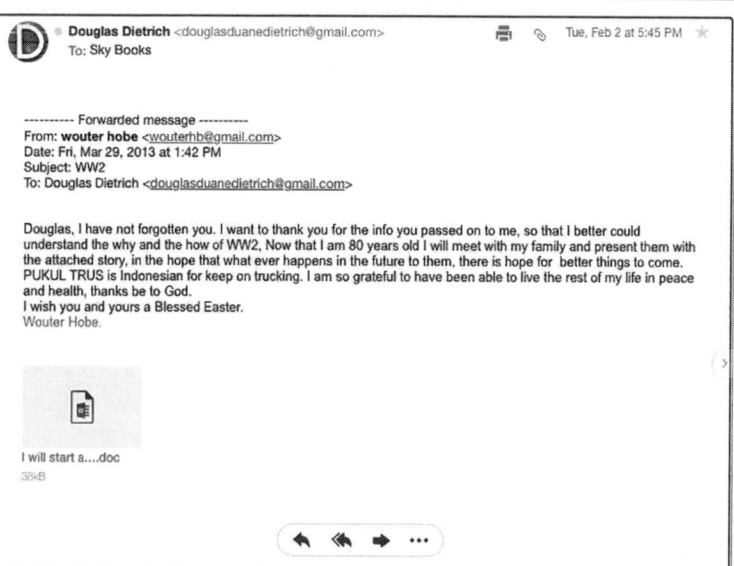

tjawet.jpg
167.9kB

Douglas Dietrich <douglasduanedietrich@gmail.com>　　　Tue, Feb 2 at 5:45 PM
To: Sky Books

---------- Forwarded message ----------
From: **wouter hobe** <wouterhb@gmail.com>
Date: Fri, Mar 29, 2013 at 1:42 PM
Subject: WW2
To: Douglas Dietrich <douglasduanedietrich@gmail.com>

Douglas, I have not forgotten you. I want to thank you for the info you passed on to me, so that I better could understand the why and the how of WW2. Now that I am 80 years old I will meet with my family and present them with the attached story, in the hope that what ever happens in the future to them, there is hope for better things to come. PUKUL TRUS is Indonesian for keep on trucking. I am so grateful to have been able to live the rest of my life in peace and health, thanks be to God.
I wish you and yours a Blessed Easter.
Wouter Hobe.

I will start a....doc
38kB

general theme of my original presentation that was presented in DVD format and entitled *Satan's Crusaders*.

Although what I have provided in this book can be considered "occult" by reason of the fact it is a hidden history, it deals primarily with ordinary facts and circumstances that were either unknown or overlooked. There were also more sinister occult factors behind the motivations for World War II which I have not even alluded to in this book save for perhaps the role of Aquino in the latter days of the Roswell mythology. World War II, however, began long before Michael Aquino was even born.

There are deeper issues to explore, and I look forward to addressing them in the next book.

BY PETER MOON

Epilogue

I would like the reading audience to appreciate that the book you have just read in a relatively short time is the result of what amounted to an eight year process of accessing the very guarded world of Douglas Dietrich and forming a working relationship with him. It was not easy and has required a considerable amount of patience. This, however, had nothing to do with Douglas's personality which I have found to be extraordinarily friendly and amenable. The best way I can describe the challenge in working with him is in a metaphorical sense. It was as if he was surrounded by fire-breathing gargoyles. A gargoyle, of course, is symbolic of a protective influence or guardian. Such a creature or function represents a double-edged sword in the most emphatic sense of the word.

Douglas's personal life is the story of someone who, by birth, was placed in an extremely unique juxtaposition with regard to major players in world history and the occult forces and threads underlying them all. What you have read in this book is only the tip of the iceberg.

First, consider that Douglas found his way by circumstances of fate where he was not only assigned to burn the most closely guarded secrets of the military industrial complex, he was also presented with crates of information about Roswell that were marked "above crypto cosmic top secret". Besides the fact that these were the actual words imprinted upon the boxes he was ordered to work with, such a labeling is over-the-top when we consider the drama it invokes. It is the stuff of which movies are made. While people might be skeptical of such claims, I can guarantee you that the stories of Douglas Dietrich's life adventures, especially when placed into a logical construct, are far more dramatic, awe inspiring and reality oriented than what you will find in an Indiana Jones movie or any other cinematic production. Even when it is not under the direct orders of the "Office of War Information", Hollywood will perpetually trivialize and misdirect by the nature of what it is.

Without even considering what might be in such boxes as Douglas was forced to confront, imagine all the frenetic excitement such a proposition, let alone an actual situation, would be generated in the minds of the multitude of zealous researchers who have pursued the UFO enigma for the last seventy plus years. Such excitement has led to demands for what is termed "DISCLOSURE", the time when the Government will come forward with "the truth about UFOs and alien contact". This, however, is like asking the Wizard of Oz to disclose the man behind the curtain. Once the little man is uncovered, he has lost the magic spell of illusion he has cast upon the people.

Another important point with regard to what has been termed Disclosure, there is no question that people will be disconcerted by the revelations of what has been considered "above crypto cosmic top secret". They will be upset by reason of the fact that it completely annihilates their preconceived notions of what they think such should or might be. It destroys all the iconography they have built up in their own minds. This, however, is what the revelation of truth should do. If it is more or less what you expected, it would be completely unworthy of being above top secret or being able to provoke such a reaction. Revelation of the truth goes hand-in-hand with ontological shock.

It cannot be overemphasized that, as he worked with these various documents, Douglas was subjected to being in the same space as Michael Aquino when the latter was conducting his occult rituals in the library at the Presidio. By the nature of what such rituals are, they cannot help but have both a subtle and profound influence on any individual in their presence. It is a bombardment of occult and magical energies that were originally written and practiced millennia ago. The fact that they were being directed with diabolical intentions makes them particularly troublesome.

In order that you might more clearly understand the challenge that I or anyone has faced in working or being associated with the "iceberg" of information that Douglas Dietrich represents, it is best to explain what is referred to in occult magick as an Egregore. The most efficient way to understand this term is via the Latin root *aggregare* which means "to gather, attach, join, include; collect, bring together," literally "bring together in a flock". This is also the etymological root of the name "Gregory". It represents an aggregate of energy which includes an assortment or potpourri of various vectors of identity, motion, influence and so on. These often manifest in the form of spirits, angels, demons, jinn or anything else within the boundaries of what might be imagined.

It is common for an Egregore to have an overall theme or purpose which can be either positive or negative, but it could also be just to create chaos. The planetary and stellar bodies and their respective archetypes represent an Egregore. To be further expository, each separate evolutionary step of life, as is mimicked in the sepiroth of the Kabbalah, represents an Egregore. For centuries, these have been represented in ancient occult manuscripts which are referred to as grimoires. The

BY PETER MOON

reason that they work is that they stir up primordial abstract energies within the DNA itself.

The Catholic mass is an excellent example of invoking an Egregore that is recognized as the "Body of Christ". This is an entire brotherhood of energies that the congregation buys into and is influenced by. Appearing to be holy from one perspective, there is also an aspect to such an Egregore which seeks to control the congregation in many ways and that includes the historical and contemporary abuses that are now so prevalent in the ecclesiastical world. An Egregore can be positive or negative, but it has been noticed by observation that most people who practice the occult will be consumed by the very Egregore they manifest.

When you consider the level of secrecy that Douglas Dietrich was working with, you had best understand that those secrets and the energies accompanying them were never meant for the common person. Part of the "Egregore of Roswell" includes the protective guardians and levels of security within the Government. The fact that Douglas's boss was so lax with security in allowing him access to such is very symbolic of the occult mechanism known as the "gatekeeper". His superior was so familiar with matters of such security that he developed a level of contempt for his own job to the point where his laziness overrode his conscientiousness with regard to doing his duty. His ability to bully through intimidation was also part of the Egregore.

On the other hand, it is also very obvious that Douglas was meant to be put in the position to access that information. His entire history predisposed him to that, and it is no coincidence that his own mother was at the crossroads of the two most powerful men in the Axis: Hirohito and Adolph Hitler. The fact that she married an American sailor and was able to emigrate to San Francisco is part of another Egregore. Her entire history fits into this very pattern, but that is too complex to be addressed herein.

With regard to more practical matters, Douglas has had to put up with continued and repeated attempts to monitor and undermine his communication. This ranges from repeated attacks on his websites to utility workmen paying undue attention to the network interface on a rather infamous telephone pole across the street from his domicile in the city of San Francisco. This even extended most recently to catching a "wandering homeless" man employing a frequency jamming device on his network. I myself have been witness to some of these cyber attacks and have even been targeted as well.

Since Douglas first appeared as a public informant, his assets were stripped and he was forced to depend upon people who were Aquino's assets who found their way to "help", offering their services for free.

Over the years, these people have made it very hard to access him, let alone work with him. To the degree that any of them were not sent by Aquino, they fall under the sway of the corresponding Egregore and were eventually consumed in one way or another.

It is therefore in this context that I explain to you why it has taken so long to access Douglas and engage in a working relationship with him. Working my way through the various Egregores, it is as if one had to pass a test. For me, it has required mostly perseverance and patience, all of which were no problem as I have a very busy schedule and many other projects to work on.

As I have alluded to already, what is offered in this book is only the tip of the iceberg. While much of the "iceberg" has already been offered online, the raw data requires a lot of context and information processing in order for it to be presented into a format where it can be digested in any meaningful way. In its most raw format, people too often irrationally hack it to bits and/or proceed to attack Douglas himself. The reason for this is that they are fighting an Egregore that is much bigger than they are and one that they are not even aware of.

One of the greatest obstacles in helping Douglas to release this information concerns Michael Aquino himself. I already knew him to be a very menacing character myself and was quite aware that he had already done everything he could to block the circulation of my own work. His influence in media was very substantial and remains so after his passing. My strategy with regard to Douglas's information was to release the book you are now reading and to do it in such a way that it would make hardly any mention of Aquino, if any at all. This was a far more effective strategy than I ever imagined as Aquino passed away from stomach cancer during the writing of this book, approximately June 29, 2020. Aquino's minions and successors are nowhere near as influential as he was. It is indeed a new era.

Accordingly, the next book will tell the story of Aquino and how he was intertwined in the Presidio sexual abuse scandals of the 1980s, all of which ultimately led to the Presidio being shut down and sold in large part to George Lucas who, in turn, sold his interest to Disney. This book will also cover how Douglas personally arranged for the arrest of Gary Hambright, the man who helped run the day care center at the Presidio and was flagrantly guilty of extensive child abuse. As this was brought to trial, the uproar and outrage by the parents of the children was so intensified that it ultimately resulted in the Presidio, the "Pentagon of the West Coast", being subject to base closure.

Accordingly, I hope you will appreciate what I say when I state that what has been presented in this book is only the tip of the proverbial iceberg. Douglas's actions in sending the Roswell debris to Japan not only resulted in international incident and the economic boom for the mainland Chinese, his actions in bringing down the Presidio made him directly responsible for the closure of the largest and most prestigious military base in the world outside of the Pentagon itself.

There is so much more to share, all of which requires time and effort, but at long last, Disclosure has been delivered. It is, however, only the first step. Do not expect the Government to come forward with any factual news on aliens or UFOs. They are, at their best, forever distracting you from the truth.

THE ROSWELL DECEPTION

APPENDIX A

THE DREAD ZEPPELIN LEGACY:
HOW THE RISE OF A GERMAN KILLING MACHINE SPAWNED THOSE ICONIC "ASIAN ALIEN SPACE SUITES"

During World War I, the Zeppelin had been responsible for the first aerial bombardment onto the capital of the Greater British Empire. This occurred in 1915 over London, England. This aero technology would prove invaluable twenty-two years later when World War II erupted, not in Poland or at Pearl Harbor, but in China. On the seventh day of the seventh month in 1937, Japanese and Chinese troops clashed outside of Beijing, and within a few days, the local conflict had escalated to total war between the Republic of China and the Japanese Empire.

Twenty-two years of technological advancement would ensue from 1915 until the lights went on mainland Asia consequent the Lugou (Marco Polo") Bridge Incident. A very important aspect of this aero technology were pressure suits, the development of which was spurred by the need to reach for ever higher altitudes from which to assail one's enemies. The pressure suits are key to understanding the Roswell deception because, as the Japanese evolved this technology, they utilized silken silvery jumpsuits for pressure protection; and, as they had shaven all of their body hair and had three fingers, they appeared to be completely unearthly to American eyewitnesses.

As this aero technology was originally German and later shared with their Japanese allies, I will give an overview of how it developed. The Zeppelin was a wholly German product. Count Ferdinand von Zeppelin, born in 1838, had been inspired to develop his airship by the sight of tethered balloons being used as observation posts during the American Civil War. After countless setbacks, he developed his first successful airship in 1900 – three years before Wilbur and Orville Wright achieved the world's first publicly recognized powered flight. It was a tremendous feat which made the count a national icon. The German Kaiser hailed him as the greatest German of the Twentieth Century. Over the next few years he improved on its design.

The individual hydrogen gas cells inside the airship's canvas envelope were better proofed against igniting the hydrogen with static electricity. Aluminum alloy produced the strength of steel at one-third of the weight, and more powerful engines gave the airships an ever-increasing range. By 1912, the airship was considered one of Germany's deadliest weapons, and the British government was at a loss as to how to combat them. Shells attached to grappling hooks were envisaged as were aerial minefields of mines dangling from balloons on a cable, of which Winston Churchill remarked : "Since Damocles there has been no such experiment."

It was actually a biplane, dropping a bomb on Tommy Terson's cabbage patch in Dover on Christmas Eve of 1914, which represented the first attack on mainland Britain for centuries. But, Zeppelins were not far behind. Once the Kaiser had agreed to bombing raids on most of London, save for his cousin's royal palaces, the Count's airships were over there with a vengeance, nigh-silently dropping death from the skies. The commander of the German Navy's Zeppelin fleet believed that Britain "could be overcome by means of airships through increasingly extensive destruction of cities, factory complexes, dockyards..." and a large number of his superiors believed him.

Many Zeppelins carried incendiary bombs bearing just such ambitions that they could generate a general conflagration. This largely failed because half of them failed to ignite and most of those that did could be easily doused; but one raid on the East End gutted several blocks of business premises, killing twenty-two people and causing half a million pounds worth of property damage. Until the middle of 1916, Britain could find no defense against Zeppelin bombers. They flew too high, and any Royal Flying Corps planes which did chase them were shot down by the airships' banks of machine guns. The balance shifted in mid-1916, however, when a member of the Brock fireworks family invented explosive tracer bullets; but the Kaiser's Second Reich struck back by inventing Zeppelins which could fly at 20,000 feet.

Despite their successes, these super-craft encountered many problems. If they were not destroyed by jet-stream winds, their crews were too frozen and deoxygenated to man the controls (see further below). By the end of 1916, any hopes of employing Zeppelins to turn London into a sea of flame had dwindled. Not until the very end of World War I did the Germans begin building vast new Zeppelins, capable of carrying a 4,000kg bomb-load. By this time, America had entered the war, and the Germans planned to drop their deadly loads over New York, thus paralyzing the will of the Americans. Of monumental consequence to history in this regard was that, after considering for twenty-four hours the ramifications of what a responsive Anglo-American Allied invasion of the German Fatherland itself would precipitate, the German Chief of Naval Operations responded with the one word : "Nein!", meaning, "no", of course.

THE DREAD ZEPPELIN LEGACY

Upon the end of World War I, the Zeppelin did not lose favor with their creators and so became the first aerial passenger transportation. It was in 1919 that the first aerial transatlantic crossing took place when a Zeppelin, flying from Edinburgh to New York against prevailing winds, arrived in 108 hours. Returning six days later, in seventy-five hours, it achieved the first double crossing of the Atlantic and paved the way for international airship travel. The huge Graf Zeppelin was built in 1928 and was named after its inventor ("Graf" is a historical title of the German nobility, usually translated as "Count" and considered to be intermediate among noble ranks; the title oft treated as equivalent to the British title of "Earl"). It carried passengers and cargo in perfect safety for one million miles, and contemporarily represented the future of long-distance air travel.

The British used rigid airships for passenger mail, but their airship program came to an end after the appalling tragedy of *R-101*, the world's biggest airship. This British industry effectively ended in 1930 with the crash of *R-101* on its way to India when it crashed in France with the loss of forty-eight out of fifty-four of those aboard. France and Italy suffered similar experiences. It is true that, across the Atlantic, American monopolized helium was nonflammable, but helium-fueled airships were also fragile and sailed into a checkered history. Of four American-built airships, two ditched in the sea and one was wrecked in violent winds.

The Germans continued with their industry, but the sabotage of the *Hindenburg* in 1937 was the beginning of the end of commercial airship travel, a service which itself had been born of a peace won via the price that the Kaiser's Kriegerflotte ("Warrior Fleet") had paid in World War I. All through The Great War, Imperial German Zeppelins attained a singular major objective in that they were able to operate well out of range of the existing antiaircraft guns and most of the Allied fighter planes. During the raids that followed, the British ofttimes could do naught but watch, grinding their teeth and knowing that their defense technology would never catch up, even with massive technological research and the expenditure of vast sums of money that they did not have in their coffers anyway. Remarkably, the American High Command would find itself echoing just such frustration(s) throughout World War II vis-à-vis the Imperial Japanese.

The high-altitude characteristics of the pioneering airships of The First World War, however, had exacted such a heavy toll on the Germans that their Samurai Allies of the Second World War had taken all necessary lessons into account by the time of their invasion of mainland China, innovating new techniques and technologies themselves prior to global war being reignited on continental Asia in 1937.

For their German forerunners, cruising at altitudes over ten thousand feet made navigation more complicated and difficult, especially when there were many moving cloud formations below. Throughout the first European war of the Twentieth Century, German aero-navigators became more and more dependent upon radio signals to stay on course, but the bearings were usually unreliable at best and any use of wireless involved signals that immediately alerted British defenders and helped the enemy zero-in on approaching raiders.

When German airships reached altitudes of thirteen thousand feet, they more frequently experienced engine trouble because of the lack of oxygen necessary to make their fuel burn efficiently. In many cases, extreme cold and icing were also problems that plagued the mechanics as they strove to keep their engines from freezing up. There was no way the power engineers could design engines that would function properly with summer afternoon temperatures that might be in the eighties at the start of the flights near ground level but would drop during the night at very high altitudes to as low as ten or twenty degrees below zero Fahrenheit.

Mechanical breakdowns affected every part of WW-I era airships and not just with the engines. The fabric became so brittle that it would break, celluloid windows shattered, control cables refused to budge, and instruments containing fluids or lubricants froze and rendered the components useless. Renovating both production and procedure enabled much less eventful patrols for the Japanese who would follow in their wake as heirs of the atmosphere.

A smoother ride was necessitated in terms of mind and body because the psychological effect on officers and men in conducting the original strategic bombing of Britain had been devastating. As many as two dozen mysterious deaths were later attributed to a form of heart failure caused by the combination of cold and lack of oxygen. Several German sky sailors died when their primitive oxygen masks failed while they were in isolated positions when manning machine gun posts atop the airship, losing consciousness long before their comrades found them and could revive them.

Altitude sickness was common and severe, sometimes rendering entire crews unfit for duty for a week or more after a high-flying raid. Dizziness, severe headaches, nausea, and vomiting were common early warning signs of this malady. On at least three occasions, Zeppelins had wandered far off course before officers onboard realized that their navigators (and in one case the commander) were only semiconscious and thus completely incapable of making any rational decisions. It was later discovered that the oils and chemicals used in the primitive breathing apparatuses issued to the crew so severely contaminated the oxygen that these victims had literally been poisoned.

A German medical officer described the confusion in the ranks of his compatriots when they were asked to determine the physiological problems and advise the Naval Airship Division on methods to counter altitude sickness. This form of illness had been so little anticipated that no research had been undertaken at that time. An account by one of the medical officers at the Ahlhorn base shed some light on the problem: "Above 13,000 feet we see signs of the now familiar altitude symptoms which manifest themselves as dizziness, ringing in the ears, and headaches. At greater heights, a marked acceleration of respiration and cardiac activity takes place. Pulse rates of 120 to 150 per minute are by no means uncommon. These severe imbalances can be controlled only by the continuous inhalation of oxygen from masks." He added some comments about the ingestion of food and liquids at high altitudes, concluding that chocolate was one of the few kinds of nourishment that could safely be consumed and that crews were better off not to eat even the lightest meals at high altitudes.

The medical researchers of the Second Reich were hard pressed to come up with antidotes for other problems induced by high altitudes such as frostbite, muscular cramps, stiff joints, and severely cracked lips. Attempts were made to design clothing that would prevent such problems, but invariably, the solutions were such that they added intolerable amounts of weight or hindered and restricted the crews from performing necessary duties in flight. Another enormous problem was fatigue, which was aggravated greatly by both the low temperatures and the thin air as well as by the fact that it was too cold for a crewman to lie in a hammock and hope to get any sleep.

THE FIRST PRESSURE SUIT

The idea for a pressure suit was outlined in 1920 by the renowned British physiologist John Scott Haldane. Haldane noted that flight above 40,000 feet would require enclosing the pilot in an airtight suit, one which would be able to maintain a proper pressure no matter what the ambient atmospheric pressure. Haldane's idea came to the attention of an American balloonist, Mark Ridge, who corresponded with its author on the construction of such a suit.

Haldane passed the letter to Robert Davis at the firm Siebe Gorman & Company Ltd (Siebe Gorman was a British company that developed diving and breathing equipment and worked on commercial diving and marine salvage projects, advertising itself as "Submarine Engineers") which adapted one of its self-contained sea diving suits for this purpose. It was tested at a pressure altitude of 90,000 feet, and it performed perfectly. 90,000 Feet (17 miles of 27 km) is the highest attainable altitude in the Earth's atmosphere – just before space. At that high level, one perceives the whole world. Any daredevil staring out from a tiny capsule lifted by a helium balloon would be looking below at the vertiginous sight that only astronauts have since observed.

No detailed records exist, but it was known to my mother that the only human subject who dared to test the British suit in the American-fueled balloon parcel in what was then deemed a clearly suicidal venture was a Japanese national who volunteered himself. Returning home immediately thereafter, he designed the highly modified rubberized (ergo shock-absorbent) asbestos-lined (ergo flame-retardant) silken silvery jumpsuits that would later be witnessed to be worn by the head-and-bodily shaven Yakuza crewmen operating the super-dirigible that crashed over Roswell.

APPENDIX B

THE SAN ANTONIO CRASH

In August of 1945, almost two years before the Roswell Incident, there was a similar crash not far from San Antonio, New Mexico. Although it has received nowhere near the amount of attention as the Roswell crash, considerable mystery and UFO drama has accrued around this event. What it was, however, was a parasite "Beetle Bomber" which had detached from one of the three super-dirigibles that were traveling to Tonopah Army Air Field. Like its Beetle Bomber parasite craft, the dirigible also crashed.

While the crash near San Antonio has been enhanced and even fabricated to fit the narrative of it being an flying saucer, reported eyewitness accounts are consistent with it being precisely what I have said. One such account is as follows:

> "As the day was edging toward dusk he was jarred from his concentration first by the feeling of an intense blast of heat followed by a deep chest shuddering air-vibration caused by a huge, weird-shaped flying object, seemingly made of metal and whining like a sick vacuum cleaner that streaked in out of the sky almost directly overhead on a slightly down-angle from parallel to the ground. The object, as it crossed out of sight barely maintaining its height advantage above the undulating canyons and rock strewn hills, all the while traveling at an ultra high speed, by the sound of it, slammed hard, and somewhat explosively so, possibly before it even hit, into the rocks and soil some distance away."

This account is from "The Wanderling", an individual whose uncle knew Dr. Lincoln La Paz, a scientist who not only claimed to have witnessed the San Antonio crash but was also brought in to evaluate the Roswell crash. Although La Paz was obviously a government disinformation agent, he clearly stated that he thought that the Roswell and San Antonio crashes were related.

La Paz even had a sighting of his own, witnessed on July 10, 1947, shortly after the Roswell crash. In a *LIFE* magazine article of April 7, 1952, La Paz was quote as saying that four members of his family had seen "a curious bright object almost motionless" near a cloud bank. It was described as follows:

"At 4:47 p.m. MST on 7/10/47, four members of the La Paz family nearly simultaneously noted "a curious bright object almost motionless" low on the western horizon, near a cloud bank. The object was described as ellipsoidal, whitish, and having sharply-outlined edges. It wobbled a bit as it hovered stationary just above the horizon, then moved upwards, passed behind clouds and re-emerged farther north in a time interval which La Paz estimated to be so short as to call for speeds in excess of conventional aircraft speeds. It passed in front of dark clouds and seemed self-luminous by contrast. It finally disappeared amongst the clouds. La Paz estimated it to be perhaps 20 miles away, judging from the clouds involved; and he put its length at perhaps 100-200 ft."

La Paz is also quoted as saying the object was huge and larger than the object that was photographed and reported on in the "Battle of Los Angeles".

The bottom line with these reports as well as all of the other reported UFO crashes in the American West, and at least 72 have been cited, they routinely conform to the narrative I have presented in this book. Is it possible that advanced alien cultures who have developed space ships are so lame or inebriated that they crash in the same general area spanning a few thousand miles?

It is self-evident that people have been so psychologically indoctrinated into the theme of an alien flying saucer crash that investigations try far too hard to fit the alien scenario into the circumstances rather than looking at the hard facts. The narrative I have presented clearly explains all the different scenarios, and it is does not require significant effort to try and make things fit. They already do. You can study this crash and find other information about it, but the point I have made is very clear.

THE SAN ANTONIO CRASH

**ABOVE IS AN ARTISTIC ILLUSTRATION OF THE
BEETLE BOMBER ATTACHED TO THE SUPER-DIRIGIBLE**

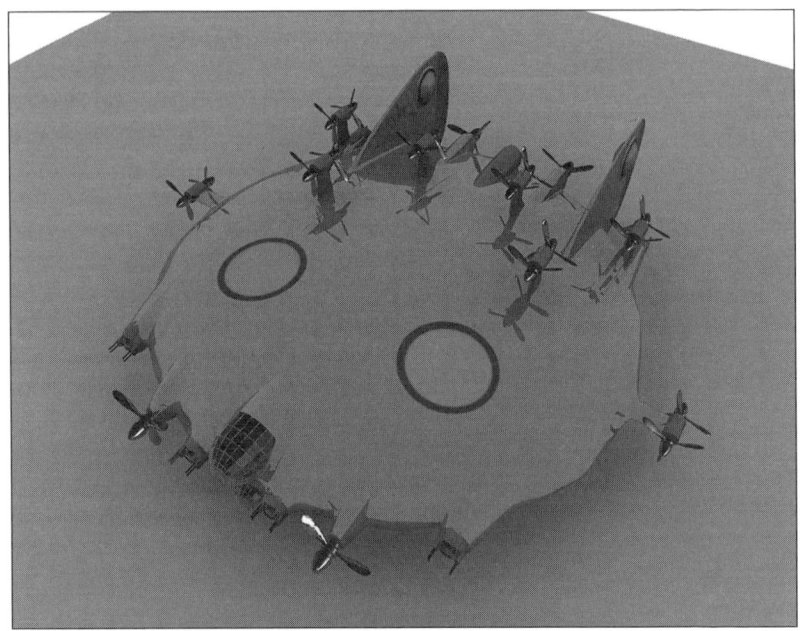

The Beetle Bomber at rest

The Beetle Bomber in flight

APPENDIX C

Attachment by Wouter Hobe

I will start about 1939. Japan was in dire straits. They needed raw materials, oil etc. The Diet made HIROHITO a God. So everything that happens was in the name of their God. Hirohito had not much to say, Tojo was in charge. He was the prime minister. They planned to invade various countries, unless they were willing to invite them for their raw materials. So they invade China. Every soldier sent overseas was prepared to die for their God. To that end they organized celebrations to say good bye to them and according to their families they have already died. So these guys were with their back against the wall. Since Japan did not have enough food, these soldiers were sent on their way without food so that they may plunder their captured countries.

America had assisted the Chinese with bombers to route the Japs, to no avail. Japan did not like that and retaliated with Pearl Harbor. Since they now needed oil, they had set their eyes on Netherlands Indies-Tarakan, in particular. But since the Dutch saw them coming, they burned all the oil. Batavia was completely in the dark from the oil clouds as all the oil tanks were set afire. So it was in Surabaja. When the Japs came into Tarakan and found out what the Dutch had done, they killed them all.

Now, here is my story. I was born in 1933 in Jogjakarta. My father was working for the government as a teacher in a trade school. But since he had done his master carpenter's diploma and he had done architectural work, he volunteered to build a part of the Petronella Hospital there and some churches for the locals. He was due for his 6 year rest in Holland, so in 1936, we went to Holland to stay with family in Doorn and in Nymegen. In 1937, we went back and now we settled in Batavia. Various family members from Holland joined us there. It was a very peaceful way of living. We were friends with the local population. We went to school there and had a good church life with choirs, etc. 7 Dec 1941 was on a Sunday; we did not go to church as we were sitting around our old Phillips radio to listen to the reports from Pearl Harbour.

My father said, "How long before they come here?"

It was in March 1942; heavy fighting in the mountains near Bandoeng. Our minister had eleven children; two of those were killed in the mountains. Two were later beheaded by the Kempei Tay. Right away, my father was out of work. We sustained ourselves, with the help of our djongos, to bake bread and make peanut butter and sell that. I was 9 years old and quite handy in roasting the peanuts. We had quite a clientele. Then the Japs came and took my father away. We followed on our bicycles but could not keep up. Later, we found out that he was in Camp Adek. We went there to bring him some supplies and saw him from a distance carrying baskets of dirt. Not long after, it was our turn. We could take some furniture, like an armoire, and beds and bedding and clothing, and we left for Camp Kramat. Everything else was left behind to be robbed by the Japs as they had nothing themselves. The gates were closed, we could not leave the camps anymore and life became harder for everybody. We had to bow down for the Japs as they were representatives of their God. I saw one tall woman getting a beating as she did not bow down deep enough. He hit her with his sword and she went down. Now, he was satisfied as she was lower than him.

Suddenly, we were told to move. This time, we could only take our clothing and bedding. We went to another prison camp called Tjideng and moved into a house already full of people. Since we moved so often, my brother and I had gotten hold of an old baby carriage and made a transport of that undercarriage. So, when the next move came we were prepared and moved much easier. This time, to another part of Tjideng. We had to be counted three times a day, standing in the hot sun for hours sometimes. At one of these occasions, there came a motorcycle with a side car that stopped not far from us. Sukarno came to do the inspection as he was now head of the police. He went into our houses and took whatever he liked.

My mother was a registered nurse and started working in the hospital. That was a school building on the main road. Our Jap camp commander was called Sonei. He had a problem. When there was a full moon, he would go into the camp with his sword drawn and would slash at anything in his way. He did not know what he was doing. Completely bonkers.

We had other Japs that had all kind of nicknames such as John the Hitter. The Japs were assisted with the Hei Ho's, Indonesian solders inducted into the Japanese army. It was unbelievable, but some people still had their dogs and chickens. We had to kill them all. I remember killing a big fighting rooster. When we had cut his neck, he was so powerful, we could not hold him and he flew into a one stone brick wall which cracked. The dogs were strung up and killed with a baseball bat. My brother was taken away to another camp; and when I became eleven years old, I was also taken away from my mother which was the most difficult time of my life.

We were transported by truck to Mangarai station and an overnight train ride into the mountains. The next day, we arrived at a station and had to be counted and walked to our new digs, a camp called Baros 6 in Tjimahi. The first two weeks I could not function and could not eat.

We were sleeping in houses with tiled floors. I had a blanket and used my clothing as a pillow. Would you believe that one tile is softer than the other? We had two adults to oversee about twenty-five boys. One of these adults told me that I had to go to the doctor who told me that I should start eating as I would not survive. Our food for breakfast was tapioca cooked in water. If you did not eat it right away, it would separate into water and something on the bottom. We also got black coffee without sugar. Lunch was a bun of bread with nothing on it.

In the camp, we had big drums stationed everywhere; where we had to pee in. This was brought to the kitchen and boiled. There is vitamin B in there. They made yeast out of it in order to bake the bread. Dinner was a tea cup of boiled rice with some veggies in it. Sometimes it was rotten Chinese potato or the green of the carrots. We supplemented it with the leaves of the Chinese potatoes. Anyway, I dried one slice of bread in the sun in order to eat that the next day and dry two slices of bread. In a week, I had an extra bun. With the rice, I did the same thing because, when we were punished, we did not get any food. We got punished for two days, so I had something to eat. The second day, however, there was nothing. That was a very painful experience.

In the meantime, I had grown a bit and would not fit anymore into the clothing that I brought along. So, I was wearing only a tjawet, a loin cloth. That was very handy as we all had chronic diarrhea, and when we had to go, we ran to the little stream that was crossing our camp. We always had to go with two people as one would look at the other's stool; and when blood was spotted, you had to report him immediately to the doctor. The guy I was with one day had blood and died a couple of days later of dysentery. The people that died were put outside, and we fashioned some sort of coffin from rolled up bamboo matting. If they died early in the evening, the next day they were already so far gone that we had to be careful not to loosen an arm or a leg. There were various camps in Tjimahi, and we found out that all these cadavers were picked up by senior people and brought to the cemetery in Lingadjati.

In order to cook for the camp kitchen, we had to get wood at the station. We had made a little pocket in our Tjawets and got some money that we put in an aspirin tube up our anus. When we got to the station, we asked the Jap to go to the little stream to do our business. It was pre-arranged with some natives that we would buy meat from them, and it was put into the little pocket in our tjawets. When we entered the prison camp again, we as little children were well endowed but the Japs never grabbed our crotches. This meat was handed into the kitchen and was directed to the sick bay for the sufferers of Beriberi. A little square of meat one inch in diameter would slink the swelling for a while. So, you understand my objections when I came to Canada and ate at a place with steak hanging on both side of my plate.

One day, there was a transport of sick people from the 9th Battalion prison camp. We mingled with the people, asking whether they knew my family. One guy told me that my brother was looking for me. We had passed each other about three times as we did not recognize each other. I was about four feet, head shaven against the lice, brown, wearing my loincloth and barefoot. My brother was tall, completely dressed, and wearing glasses. He told me that he had brought my uncle Pieter there as he was suffering from Beriberi. Now, I knew where my brother was.

I had made friends with the guys working the kitchen. They always had something extra to eat which they shared with us. One day, there was a cat walking around, but not for long, as the next day we got a piece of meat from them. I saw one day a bunch of guys standing opposite each other and slapping each other in the face. The Jap stood there with his sword, ready to "encourage" them.

After the war, I started to work for KLM. Once there was a camp reunion in Driebergen, and I saw my colleague from KLM there. I asked him about the camp and he said that he was involved in that beating. His one eye was crooked in the socket.

When the war was over, we only heard that about September 22, 1945. I was called to go to my locker. There was a guy standing there. He asked me whether I was Wouter Hobe. I said, "Yes!" He said that he was my father. I did not recognize him. He said to come with him. I refused, but he had made arrangements already with the camp commander and told me that my brother was waiting outside. So, I came along. I had no more clothing as I had traded that with the local population for a comb of bananas. That stopped my diarrhea for good. So, we started walking from Tjimahi to Bandung where my father's camp was situated. It was a very dangerous thing as the local population was now riled up by Sukarno and his gangs. We stopped at a Japanese lorry and they brought us to the Camp 10e Bat. in Bandung. Through all the transports between camps, my father had found out that my mother was still alive and working in the hospital barefoot. So, he made a pair of kleteks (klompen). I burned her name on them and they were sent with the next transport to Batavia (they are now in the museum in Arnhem called Bronbeek).

We were sleeping in a large hall. There were a lot of bed bugs in the mattresses. My father had made bamboo beds, chairs and a table and everyday we had to bring them outside into the sun and stamp them on the ground in order to remove these bugs which fell out in big lumps. I slept with my mouth open one day and one crept in. I woke up, closed my mouth and crushed one that was sitting on my molar. To this day, I cannot eat any cilantro as it is the same smell. Ask any totok about kutu busuk and they will confirm that.

My father was disqualified for his health and we managed to recuperate in Australia. The red Cross saw to it that my mother was also informed and made her arrangements. She told her patients that she was leaving. One lady asked her to go to Perth and visit her sister there. We were transported by train to Batavia under heavy Ghurkha guards. These Indonesians were very much afraid of these Ghurkhas with their shaven heads except for one strand of hair. They had a rifle in their left hand and a sword in their right hand, and they were regarded as devils.

We stayed overnight in a hotel and had to sleep behind sandbags as the whole night there was fighting in the streets. The next day, we were transported by army trucks to the harbor, and my brother spotted my mother first. We got united and boarded a Japanese freighter who brought us outside the harbor where the ship *Oranje* was waiting. She could not enter the harbor as a ship was sunk in the entrance. When we got on board the *Oranje*, it was lunch time and we could look in on the restaurant where everybody was eating potatoes, carrots and fish. It must have been a Friday. When it was our turn to go and eat, we were seated at tables with a white linen cloth. There was butter and jam and cheese. Then, the bread was brought in. Huge slices of white bread. We all stared at it. Then, somebody said eat and everybody took a slice and dug his face in it to smell. We ate it as cake but could not eat any cheese or butter. It was too fat for us.

When we arrived in Freemantle, we were boarded on a train to Perth. It was a Sunday. There were no people in the street as everybody was at the beach. We walked single file to the hotel Wentworth. After we were settled, we walked about. We passed a green grocer who had done up his window with fruit and vegetables, and we stood there in awe, looking AT HEAVEN.

A couple of days later, we were visiting with the family of my mother's patient, and a girl opened the door and ran inside calling for her mother. We made friends. After nine months there, we were going to Holland. We were told not to mention that we had family in Holland. The trip was very difficult for me as I had broken my arm while traveling in the Red Sea and was in the hospital when we went through the Suez Canal. Upon arrival in the Hook of Holland, we saw people standing on the pier with banners. Somebody had binoculars and could read what was on them. "Rice pickers go back." Anyway, the ship had to maneuver quite a bit around all the cranes that were toppled in the harbor, and when we attached to the "Kay", we heard the Dutch national anthem being played. Upon our disembarking, Queen Wilhelmina was standing there and welcomed us to Holland. She gave us a scroll that spelled out that the Dutch Government was going to do everything in their power to integrate us into the Dutch society. But since the government was socialistic, nothing came of those promises.

I finished my education and had to join the army. I was out within a year as I got meningitis. Then, I started to work for KLM, and after 3 years, I emigrated to Canada where I worked at Dorval Airport, first for BOAC, then KLM, and then CP air. When I was with CP for twenty-five years, my wife and I went to Australia to have a look around. We visited Perth and that little girl from 1946 was now an adult and married. The old mother had Alzheimer's and could not remember anything. But, this girl told us that she was so frightened when she opened that door as she saw a ghost standing there. Anyway, she had family in Holland and her cousin was working for KLM. He was in a college, and I worked with him for three years.

Thank you.

THE MONTAUK PROJECT
EXPERIMENTS IN TIME

SILVER ANNIVERSARY EDITION

A BRAND NEW VERSION

The Montauk Project was originally released in 1992, causing an uproar and shocking the scientific, academic, and journalistic communities, all of whom were very slow to catch on to the secret world that lurks beyond the superficial veneer of American civilization.

A colloquial name for secret experiments that took place at Montauk Point's Camp Hero, the Montauk Project represented the apex of extensive research carried on after World War II; and, in particular, as a result of the phenomena encountered during the Philadelphia Experiment of 1943 when the United States Navy attempted to achieve radar invisibility.

ISBN 978-1-937859-21-3 $22.00

The Montauk Project attempted to study why and how human beings, when exposed to high powered electromagnetic waves, suffered mental disorientation, physical dissolution or even death. A further ramification of this phenomena is that such electromagnetic waves rescrambled components of the material universe itself. According to reports, this research not only included successful attempts to manipulate matter and energy but also time itself.

It has now been over twenty-five years since *The Montauk Project* originally appeared in print. In this *Silver Anniversary Edition*, you will not only read the original text, accompanied by commentary which includes details that could not be published at the original time of publication, but also an extensive summary of a twenty-five year investigation of the Montauk Project which culminated in actual scientific proof of time travel capabilities.

ORDER TODAY FROM SKY BOOKS

Spandau Mystery - by Peter Moon

The end of World War II precipitated more intrigue and struggle for power than the war itself. Much of this centered around the secret projects sponsored by Rudolph Hess which included not only the Antarctic project but the construction of Vril flying saucers. These tasks eventually crossed the path of one of the most colorful characters of the Second World War: General George S. Patton. Patton's job, as the war came to a close, was to recover the secret technology of the Germans and safeguard it for American use. After accomplishing his mission and compiling a German history of the war, General Patton was killed in a dubious accident, the mystery of which has never been solved and has been magnified by government refusal to declassify the file on the investigation of his death. Far more conspicuous and powerful than Patton was Rudolph Hess, the Deputy Fuhrer of Germany, who flew to England in 1941 as an envoy of peace and was imprisoned for life and suspiciously killed just before his imminent release. The current of intrigue and power which permeated these two individuals and led to their downfall was the same current which led to a repatriation of the U.S. Government and an undermining of a constitutional government that is run by and for the people. It was thus that Patton and Hess wore different uniforms but shared common interests and held within their grasp a force so powerful that it resulted in murder for both. This is a novel by Peter Moon which led him to the real life revelations of Douglas Dietrich.
350+pages, ISBN 978-0-9678162-4-1.....................................$22.00

SkyBooks

There is an order form on the back of this book if you would like to purchase the above book, and if you would like more information on these or additional titles, you can visit the websites below.

WEBSITES

book store: www.skybooksusa.com

www.digitalmontauk.com

www.timetraveleducationcenter.com

THE MONTAUK PULSE

The *Montauk Pulse* originally went into print in the winter of 1993 to chronicle the events and discoveries regarding the ongoing investigation of the Montauk Project by Preston Nichols and Peter Moon. It has remained in print and been issued quarterly ever since. With a minimum of six pages and a distinct identity of its own, the *Pulse* has expanded to not only chronicle the developments concerning the Montauk investigation, but has expanded to include all the adventures that have surrounded Peter Moon since that time. This includes his adventures with David Anderson and the ground-breaking events that are occurring in Romania. It also includes his interaction and updates concerning Douglas Dietrich. Subscribing to the *Pulse* contributes to the efforts of the author in writing more books and chronicling the effort to understand time and all of its components. Past support has been crucial to what has developed thus far. We appreciate your support in helping to unravel various mysteries of Earth-based and non-Earth-based consciousness. It makes a difference. You can subscribe for $20.00 annually if you are in the U.S.A. or $30.00 if you are overseas. See the order form on the back of this page.

The Time Travel Education Center

The Time Travel Education Center was created in 2015 in order to educate the public on the simple math and science behind the concept of time travel (with free videos) and also to keep people informed on related aspects to this very avant-garde and rarified subject. The science and math, based upon the genius of Dr. David Anderson, are introduced at an eighth grade level of mathematics yet the concepts are astonishingly profound.

Peter Moon has also prepared an on-going video series on the **Psychology of Space-Time** in order to help people understand the issues surrounding this phenomenal technology and why it is not readily available for everyone. There will be further videos as time allows.

You can become a free member of **The Time Travel Education Center** by going to the website below, and you can also become a paid subscriber which will give you access to further information including books in progress by Peter Moon. Your support is important.

VISIT THE TIME TRAVEL RESEARCH CENTER:

www.timetraveleducationcenter.com

Sky Books   ORDER FORM

We wait for ALL checks to clear before shipping. This includes Priority Mail orders. If you want to speed delivery time, please send a U.S. Money Order or use MasterCard or Visa. Those orders will be shipped right away. Complete this order form and send with payment or credit card information to:
Sky Books, Box 769, Westbury, New York 11590-0104

Name
Address
City
State / Country   Zip
Daytime Phone (In case we have a question) (   )

☐ This is my first order   ☐ I have ordered before   ☐ This is a new address
Method of Payment: ☐ Visa   ☐ MasterCard   ☐ Money Order   ☐ Check
—   —   —
Expiration Date   Signature

TITLE	QTY	PRICE
The Montauk Pulse (1 year - free shipping US orders)...$20.00		
The Montauk Pulse International (1 year)..................$30.00		
Montauk Project SILVER ANNIVERSARY EDITION...$22.00		
Spandau Mystery...$22.00		
Note: There is no additonal shipping for the Montauk Pulse. International subscription is $30.00. **Subtotal**		
For delivery in NY add 8.625% tax		
U.S. Shipping: $5.00 for 1st book plus $1.00 for 2nd, etc.		
Foreign shipping: $20 for 3 books		
Total		

Thank you for your order. We appreciate your business.